MW00333736

"A must-read for any executive who's looking to understand what's next and get ahead of it."
Claire Valoti, former VP EMEA, Snapchat Inc. and Head of Agency Relations, Meta

"Concise, actionable and insightful, *Disruptive Technologies* will help executives breathe easier and sleep better."
Daniel H. Pink, author of *Drive* and *To Sell is Human*

"A comprehensive look at the fast-approaching technology tsunamis guaranteed to change your business. This book is an insightful and practical guide to making decisions in an uncertain world."
Nir Eyal, author of *Hooked: How to build habit-forming technology*

"Some books have ideas, some vision, some practical advice. This is one of the rare books that has all three. A good read for anyone who actually wants to do something about tech change and not just talk about it."
Marco Rimini, CEO, Worldwide Central Team, Mindshare

"The perfect guide for all professionals looking to better understand and thrive in the current marketplace of disruptive technologies."
Michael Villaseñor, Director of Product Design, Spotify

"An integral how-to guide for the future of your business. If you want to disrupt a market (or avoid being disrupted) you need this book."
Chris DeWolfe, CEO, Jam City

"If you want to embrace change rather than become its victim, read this book. Highly recommended."
Richard Watson, author of *Digital vs Human*

"This book provides a practical blueprint for dealing with change and disruption in the age of abundance. From adopting a mindset for disruption to organizing a product brainstorm, it's a one-stop shop for ambitious today's entrepreneurs."
Russ Shaw CBE, Founder of Global Tech Advocates

"*Disruptive Technologies is* an elucidating look at the technology that is likely just around the corner. Paul is full of advice on how to figure out and prepare for what might be ahead, and set us up to ask the right questions about what tomorrow will look like."
Mike Murphy, Editor-in-Chief, IBM Research

"Paul's experience, energy, honesty and insights make *Disruptive Technologies* a must-read."
Reshma Sohoni, Founding Partner, Seedcamp

"An informed, concise and above all practical manual to help you successfully navigate disruption."
Shawn Gold, CEO, Pilgrim Soul, and formerly Adviser to Executive Council, TechStyle Fashion Group

"At a time when innovation is proving make or break for so many firms, it's great to see someone with Paul's knowledge of technology tackle the topic."
William Higham, Founder and CEO, Next Big Thing

"I'm tracking a lot of what Paul is covering here in a super-readable and actionable format. This book needs your attention. Dig in now."
Chris Brogan, CEO and Owner, Media Group, and *New York Times* bestselling author of *The Freaks Shall Inherit the Earth*

"The future is already here – we just haven't noticed yet. Paul's book is a crucial guide to the immediate future, post-disruption, and powerfully directs your attention towards the key opportunities. A must-read!"
Gerd Leonhard, futurist, author, and CEO of The Futures Agency

"As the volume of the conversation around disruptive technologies increases, it becomes harder to separate the signal from the noise. Paul Armstrong's approach is both accessible and practical, enabling anyone facing disruption (i.e. all of us) to better understand how they might respond. It's a great read for anyone wanting a primer in how to navigate tomorrow's world."
Henry Coutinho-Mason, author, *The Future Normal* and *Trend-Driven Innovation*

"Lots of books attempt to predict the future of technology. This one does something different and more valuable: it provides thoughtful, step-by-step advice on how companies can prepare themselves for the era of unprecedented change that lies ahead."
Harry McCracken, Technology Editor, Fast Company

"Paul provides a good overview of how technological disruption happens, and how you or your company can make it happen for you. Clear and concise."
Mathew Ingram, Chief Digital Writer, *Columbia Journalism Review*

Disruptive Technologies

A framework to understand, evaluate and respond to digital disruption

SECOND EDITION

Paul Armstrong

KoganPage

First published in Great Britain and the United States in 2017 by Kogan Page Limited
Second edition 2023

2nd Floor, 45 Gee Street
London
EC1V 3RS
United Kingdom
www.koganpage.com

8 W 38th Street, Suite 902
New York, NY 10018
USA

4737/23 Ansari Road
Daryaganj
New Delhi 110002
India

Kogan Page books are printed on paper from sustainable forests.

ISBNs

Hardback   978 1 3986 0922 8
Paperback 978 1 3986 0920 4
Ebook   978 1 3986 0921 1

British Library Cataloguing-in-Publication Data

A CIP record for this book is available from the British Library.

Library of Congress Control Number
2022052514

Typeset by Integra Software Services, Pondicherry
Print production managed by Jellyfish
Printed and bound by CPI Group (UK) Ltd, Croydon, CR0 4YY

CONTENTS

ACKNOWLEDGEMENTS

First and foremost I would like to thank my family: my mother, who taught me the importance of empathy and authenticity, my father, who taught me right from wrong and encouraged me to be bold, and my sister, who (despite me being an utter rotter to her while growing up) is my greatest cheerleader.

I am also grateful to my other 'family'; the friends who are also a constant source of inspiration, strength and good times. The university family: Anita, Collette, Dylan, Emma, Jennie, Jess, Joanne, Nicola and Peggy. Not forgetting the Los Angeles family: Aimee, Carla, Giovanna, Mike, Jess, Mark, Jay, Vince, Corey, Marla, Jen and Phil. I thank you all for your support over the years and am humbled by your loyalty.

Disruptive Technologies would have been much harder to write without the advice, assistance and encouragement given by Darika Ahrens. You are a constant source of inspiration who crystallized a lot of my thinking and helped me push things further.

I would like to thank Kogan Page for approaching me and the following 'north star' people who have helped me along the way – each a star in their own right but also amazing teachers, friends and sources of inspiration: *Lisa Becker, Josh Brooks, Rhonda Brauer, Matt Broom, Jonalyn Morris Busam, Brenda Ciccone, Suze Cooper, Cristian Cussen, Robbie Daw, Carolyn Dealey, James Denman, Emma Diskin, Kevin Dunckley, Robinne Eller, Lisa Fields, Georgina Goode, Emily Hallford, Dave Halperin, Lisette Howlett, Emily Kealey, Jason Kirk, Michael Levine, Deborah Peters, Brianne Pins, Nicole Randall, Claire Selby, Simon Speller, Craig Stead*, and the many others who know who they are.

Finally, I'd like to thank you, the reader, because without people like you – people who are interested in things like the future, disruption and doing things differently – the world would be a in lot more trouble.

Thank you all.

Introduction

Stop for a second and imagine a world with no social media, where a Bitcoin would only hurt your teeth, TikTok was just the noise a clock made, Uber only meant 'above' in German, people would have thought you were bad at spelling if you wrote Canva, Alibaba was just an unlucky woodcutter, people couldn't use their phones for much more than making a call, commercial space travel was only in science fiction comics, no one knew what an iPhone was, 9/11 hadn't happened, the only financial crises talked about were in history books, no one knew what COVID-19 was or who Edward Snowden was and America had only had Caucasian presidents.

This was the world 20 years ago.

Fast forward to today and all these things either feel old or very much 'within our lifetime'. When you think about the above list of changes and evolutions it becomes clear that these did not happen for reasons of politics, population shifts or natural selection as seen previously in history. These changes were calculated to fix problems – real or perceived – and many had no physical element to them. Technology has been the great transformer here, be it the internet, increased collaboration or new 'needs' emerging. Technology has been the driving force of change and will continue to be for decades to come because of its proliferation into our everyday lives and more importantly behind the scenes. Indeed, to previous generations it could sound hyperbolic to say that a vast amount of what is normal today will seem ludicrous to those looking back in 20 or 30 years' time, but few would argue this is the case today if we look back over the past 20 years. The rate of change this generation is seeing is unprecedented. Whereas previous decades were fraught with wars the likes of which had never been seen before, war for this generation and subsequent ones will be an ongoing and regular occurrence broadcast in high definition straight to their mobile device – a sad reflection of modern life.

How and why has technology advanced so quickly?

Understanding this question is pivotal to understanding where technology is going and how technologies will change further still. Improved communication has brought about much of the innovation and technological change we see today, whereas previous decades and generations did not have the means to collaborate so freely or widely. However, greater communication is not the only thing that has brought about the rapid change we see today:

- **Advancements in processor power**
  The processor power in the average smartphone now eclipses the functionality of the early floor-filling sprawling towers that were the first computers. We can also do more with the devices in the palms of our hands than we ever could with those massive, ancient wires. Make no mistake, though; without them and the learnings that came with them, the smartphone, PC and other industries would be in a very different state. Interestingly, we may be on the cusp of a leap into new areas if Moore's Law – the principle that every two years the processing power of computers doubles – is not as infinite as it was once thought to be. While the law has been a constant for several decades (Intel, 2015) and will likely be for at least the next decade, its future – and what comes after it – is still being debated.

- **Miniaturization of materials**
  This has enabled smaller parts to be made and has given us the power to do more using far fewer materials. This has driven costs of materials and creating products (especially electronics) further and further down. Not only has this saved money for manufacturers but it has also meant changes to the way things are made – often with a lot less energy. While very few technologies could be dubbed green initiatives, it means that what was once an enormous environmental polluter for the world is now a lot less harmful (although this is far from having no effect on the environment).

- **Rapid prototyping**
  The ability to create a scale or full-scale product swiftly and cheaply, using the computer-aided design techniques that were born in the 1980s, heralded a new era for technology. Parts and full products could now be created and tested, whereas previously full batches had to be produced. Rapid prototyping and the early rectification of flaws have led to higher successful potential of products.

- **Increased connectivity**
  More connections to the right people and an infrastructure to share and promote knowledge (the internet) have enabled significant leaps to be made in various technological areas. This has gone beyond simple computing and has extended to education, business and transportation. The ability to observe and ask questions in order to continually improve best practices and ways of doing things has fundamentally shifted companies and countries to new levels. Instead of simply being able to utilize the knowledge of your immediate peers you can access a vast array of intelligence and perspectives enabling greater discoveries to be made, faster and with less risk (be that wasting time, money or another resource).

- **Lower cost of storage**
  The cost of physical storage has plummeted, enabling new options, including cloud storage, to come into the public realm. Essentially, storing files and manipulating them on a distributed network – often referred to as the Cloud – has saved money on physical materials and removes additional laborious processes.

Where will technology go next?

Technology is becoming more and more amorphous. What once was a simple-to-understand spectrum has become a difficult beast to wrangle. Clients often find a shoal of fish analogy works well. Fish will come together to fend off attacks from larger prey, forming a larger mass to scare away predators. When looking at technology it can be like this – a larger entity that is made up of many smaller entities, each following a similar path, but when pressure gets too much, change occurs and the shape reforms.

The changes that are happening are like grabbing a balloon – if you squeeze one end too tightly, it will burst out in another area. Some of these bursts can be predicted, others cannot. The ability to predict such movements comes from careful analysis of a variety of criteria, knowledge, the landscape and often a lot of other data. Other movements cannot be predicted and these changes often disrupt the landscape much more than the predictable movements of various fields.

Years ago, technology was simpler because systems were not connected and computers and gadgets could only perform much more basic tasks. Those days are far behind us and while it is unlikely we will ever truly lose 'dumb' technology (after all it was made to make lives easier) the future is exciting, with connected systems holding the potential for making things better (with some equally serious potential drawbacks, discussed in later chapters).

Connecting systems is an exciting arena. Due to such advancements – especially cloud computing – a new future, the Internet of Things (IoT), is on the horizon. First coined by Procter & Gamble's Kevin Aster in 1999, IoT is essentially a world full of cheap sensors deployed in everything from toasters to train tracks that all feed into systems that collate and analyse 'big data' – a term first used by John Mashey (Diebold, 2012). From smart cities to predictive modelling, big data is a huge part of the future for every person on the planet. Sensors in things are nothing new but the ability of connected systems to solve larger problems faster and cheaper is, many believe, what will shape us for many decades to come.

The world is about to get a lot more complex

Based on McKinsey data, the global market for the IoT will enable $5.5 trillion to $12.6 trillion of value to be unlocked globally by 2030. A huge sum of money that includes value from consumers and customers of IoT products and services. The developed world will account for 55 per cent (decreasing from 61 per cent in 2020) with China poised to become the dominant global IoT force due to the country's dual manufacturing and technology supplier capabilities along with a huge end market to cater to.

The key point behind the IoT is the aforementioned big data element. Previously, it had been possible to collect relatively large sums of data but visualization and analysis were difficult. Improvements in the statistical and computational analysis of data sets (along with the exponential increases in computer processing power) now mean that such shackles have been removed. Data investigators (also known as 'data scientists') are now some-what freer to begin finding new patterns and areas to achieve different objectives, whether it is saving money by limiting production at certain times based on demand, redirecting traffic to smooth out traffic jams, or helping diagnose mental illnesses based on chromosome patterns. The potential for big data to change the world is clear although making changes based on the results remains more complex than is often described.

Huge corporates aren't the only ones who benefit from big data. In fact, it may just be governments or city departments who gain the most value in this area due to their ability to take time making changes, or else having multiple systems at their disposal so changes can happen fast. While big data certainly has the potential to help companies leapfrog competitors, it is the combination of data, visualization and analysis working together that will offer businesses the most significant gains. The list of what big data can do is often bewildering and it is a growing area – a quick Google search throws up more than one billion, four hundred and eighty million results for 'Degrees in Data Science', up from 26 million in November 2015. Big data is not easy and requires a specific skillset (for now at least) but more so it requires the right questions to be asked or the wrong answers will be found – data is only as good as the analysis and insights drawn from it. Beyond this, there are other issues with big data that fall into a couple of areas – speed and being correct. Big data gets the name because it deals with lots (sometimes billions) of bits of data, and apart from being cumbersome to manipulate, finding real value from this complexity can be difficult because of the size involved. Beyond this, the amount of data can lead to false positives if not handled correctly – the complete opposite of the desired outcome. Quick wins and short-term goals are often sought by businesses grappling with big data to demonstrate success and recoup costs, but big data requires care, time and attention in order for maximum value to be achieved.

Understanding big data (let alone acting on it) is not easy. Senior VP and Chief Learning Officer at software experts SAP, Jenny Dearborn, explains it succinctly (Dearborn, 2015):

> Once you understand the deeper processes that underlie your data (descriptive analytics), have a sense of why they are happening (diagnostic analytics), and have predictions about the future (prescriptive analytics), the next step is to act on your knowledge.

The ability to predict the future is of course incredibly desirable for a large number of reasons and while the aim of this book is not to predict the future it is most certainly to help businesses and individuals plan for it. To plan for something, you need to do three things: one, understand it ('What is this thing?'); two, evaluate it ('How important is this thing to me?'); and finally, you need to respond, even if the response is to do nothing ('What should I do about this?'). Each step is critical, and this book deals with each in turn, specifically enabling you to not only understand but to evaluate and respond to that area (and change) accordingly.

Prediction is hard but that shouldn't stop you

Business leaders understand that prediction is part of any business that needs to improve – adapting to changing times and technologies is far easier to say than do. Predictions can be put on a scale – simple or complicated, based on the data you have to hand. Everyday life prediction (the state of traffic at different times, when the postman will come) can be reasonably accurate based on previous patterns. But what about whether one country will invade another country in 10 years' time and the resulting economic effects on a third country's iron ore industry in 15 years' time? This type of prediction is hard to do and no amount of historical data, previous experience or gut feeling will suffice because the data would be insufficient as a basis for real decisions.

People and industry rely on individuals and collections of predictors (think tanks, trend agencies) in the absence of any legitimate accurate data or collected 'database of truth', which is all well and good until you realize it is impossible to predict the future with so many unknown variables and multiple moving parts. Studies have shown (Tetlock, 2015) that even the best 'experts' (a full collection of academics, pundits, consultants and specialists) may be no more accurate than a monkey with a dartboard over a 20-year period. The results of the Tetlock study showed that for shorter periods (three to five years out), some experts were significantly better than others but most participants floundered (or results were significantly poorer) the further out they were asked to predict (five or more years). The results of this study are why this book has been written, focused and structured in the way it has. This does not mean that trying to answer the questions is futile or that we shouldn't try to do difficult things, but it does mean we need new tools, systems and ways of thinking in order to solve such problems. Previously, business used static models that were inflexible but were fit for purpose. However, in today's changeable world and technological ecosystems, such models are no longer suitable. It is for this reason that a flexible framework is required and TBD (Technology, Behaviour and Data) was created. The framework focuses on the heart of decisions that need to be made today, tomorrow and in the future. Prediction is hard but rolling with the punches can be easier if you have a workable system that enables you to make choices and not be led by those forced upon you.

Any good technology has people at the core of it

Technology has always been a force in the world; whether your job was the distribution of the Bible, or moving iron ore from one place to another, technology was there to make things easier, smarter and faster. In today's world (and the future), technology is again taking on a different role, as physical tasks are completed by robotic devices and mechanical processes become invisible in our everyday lives. Instead, we see the function of technology evolving to deal less and less with physical needs, somewhat in line with Maslow's Hierarchy of Needs (Maslow, 1943). This theory consists of five key stages of human motivation: physiological (air, water, food, heat and so on), safety (protection from the elements, physical security, laws), social (intimacy, love, affection), esteem (achievement, independence, prestige, self-respect) and self-actualization (personal potential, self-fulfilment). While Maslow has been critiqued since he introduced the model (and other versions have been put forward) it is the humorous ones that stay in people's minds – for example, one that adds another layer on the bottom of the pyramid entitled 'WiFi'. While comical, this simple addition does show the pervasiveness of technology in people's everyday lives and why technology can become as much a disabler as an enabler.

Therefore, it is important that any strategy or framework has an element that corresponds to how humans will interact or be impacted by any technological change. Any new system you create will elicit behaviours from those that are affected by it, so it is important to look at existing behaviours before, during and after any change to make sure that you are making the right choices and any desired outcomes are realized with the fewest adverse outcomes. Understanding this impact is a major area of focus for this book, as any change requires careful handling for a number of reasons but mainly because humans are weak at dealing with change, and technology (somewhat) has the role of replacing or reducing the necessity for humans in a lot of industries. A good example of this would be optical character recognition software allowing computers to sort mail with very little human oversight. The mail was a simple system that was completely revolutionized forever with a small technological change. Fast forward 50 years and Meta has baked artificial intelligence helpers into most of its products so that you can book hotels, flights and theatre tickets and even recommend ways to cheer you up. The reverse of the changes in the postal system is now happening; large changes are coming to simple systems, offering gargantuan potential outcomes. Meta has completed some incredible technological feats (not without some large mis-steps); none, however, is without issue when you think of the industries the company continues to disrupt.

A new breed of technologies is on the horizon; a clear and decisive set with not only the potential to move minds and markets but perhaps the future of our genus if we are to colonize other worlds someday. These technologies are dubbed 'emerging technologies' and as we explore these in more detail we need to think about the people who will use them. Arthur C Clarke famously once said, 'Any sufficiently advanced technology is indistinguishable from magic', and this can be said for a lot of the emerging technology we'll discuss in later chapters. Overall, the public remain cautious (but optimistic) when new technologies arrive. Often it is the way that such technologies enter our lives or are introduced that has the greatest impact on whether they take off or wither. This is why exploring the people (or behavioural) aspects of any technology (and change) is so important. When jobs or liberty are threatened, people react badly towards technology. This 'effect' can be seen throughout Hollywood depictions of technologies in *Minority Report*, *A.I. Artificial Intelligence*, *Ready Player One*, *I, Robot* and beyond. This said, it is your job (and mine) to push these comfort zones, sometimes probably without asking for permission. Despite there being no firm reference for where or when he said it (if he said it at all) Henry Ford's famous quote sums up why: 'If I had asked people what they wanted, they would have said faster horses.'

Adaptability for the win

Blockbuster, Enron, Woolworths, Borders, Comet, Jessops and Oddbins. All decent businesses at one time but now consigned to the pages of history and Wikipedia. Twenty years ago, many thought these companies were untouchable and yet today they have all ceased trading and seminars are run on how not to be the next [insert failed business name here]. These failures aren't one-offs or victims of sad circumstances – they are, for the most part, examples of companies that floundered in the face (or perhaps in some cases, pace) of change. Peter Diamandis, Chairman and CEO of the XPRIZE Foundation, best known for its $10 million Ansari XPRIZE for private spaceflight, famously quoted a statistic from a Washington University study: '40 per cent of today's Fortune 500 companies will be gone in the next 10 years' (Ioannou, 2014).

This quote was one of the founding thoughts behind this book. To put it another way, if the statistic holds true (and there's no reason to suggest it won't, and it may even speed up), every 3,650 days there will be 40 per cent turnover of the major corporations in the world. Who will they be? Will yours be one? What will the new ones look like? How will they impact your business? How will they impact your life?

When you read headlines and statistics like this, it's easy to think that these changes happen because the companies have bad products, poor leadership or challenging economic circumstances. However, the truth often runs much deeper. These businesses often lack the foresight and/or the willingness to adapt to changing times, whether related to technological changes, behavioural changes or changes to do with data and the way it is utilized. These elements are core throughout this book; the technology itself, the people who will use or be impacted by the technology and the data that will be required, created or changed because of the technology and changes occurring. Understanding the why behind all these questions is imperative if we are to go beyond surface implications – behaviour is a key one as, ultimately, any change will impact humans – indirectly or directly.

Business isn't getting any easier. Despite the fact that it is easier than ever to start a business and that we have access to new tools that previous generations could only have dreamed of, many are struggling in this emerging 'new world'. The one certainty is that not all will survive. Many businesses can survive one 'hit' from technological (the focus of this book), political, social or economic factors but it is because of the various forces currently colliding that many are seeing their boats fully overturned almost overnight or (possibly worse) over time. The one constant, to paraphrase Greek philosopher Heraclitus, is change. Those companies that can adapt to changes happening around them, and quickly, stand the best chance of future and sustained success as well as the lowest risk of major disruption from outside forces as their ability to flex with the market and competitors' movements is that much more robust.

Having a system is key

Businesses and individuals need better systems to understand the changing world around them. The first step is to understand what it is you are dealing with, the second is to evaluate the thing you are looking to have an impact on and the third is to act on the first two elements with an appropriate response based on your aims, goals and objectives. Simplicity is then the key to making the first practical steps to change.

In Chapter 1, you will be introduced to emerging technologies and upskilled into a subset of technology that has the exponential potential to genuinely transform businesses (and in some cases humankind). The chapter will look at how each of a carefully selected group of technologies will impact business and how you should think about them as they mature and morph into new technologies and opportunities for your business. Chapter 2 discusses the disruption that surrounds these technologies and how they and others are often misunderstood before their value is fully appreciated. This chapter also explores how applying technology poorly is damaging your customer relationships. Following on from this, Chapter 3 looks at the 'innovation expectation gap', and details how innovation happens and why it is important. Chapters 4 through 7 detail the flexible 'TBD' framework (technology, behaviour and data) and change evaluation system and discuss why the flexible nature of the system is required for long-term success. Chapter 8 is about the application of the TBD system and will guide you through using the TBD when it comes to business decisions via a step-by-step guide. By the end of Chapter 8 you will be able to decide whether to invest in or ignore changes and will be armed with practical tips for implementation of the TBD framework in the future. Chapter 9 will look at 'dis-innovation', why thinking differently won't save you and how to make the right changes now so you are ready for 2026 and beyond. Chapters 10 and 11 investigate and help you to evaluate emerging technologies surrounding the future of the web, namely Web3 and the multiverse. Chapter 12 looks at why the Millennial generation is so important to the future, how 'Millenovation' is the next step in your company's future and how to utilize the Millennial resource rather than being swept away by it. The final chapter will look at the future of TBD, emerging technologies and how to make sure you aren't overwhelmed by the next big thing.

Change is hard. Don't let anyone tell you differently. While it is easy to say, it is most definitely not easy to do – especially in a corporate environment where there is no finish line. This book, and more precisely the TBD framework, will give you confidence and prepare you to take a similar journey multiple times.

01

Emerging technologies

The world is certainly a smaller, more intricately connected place because of the digital revolution and one also still very much in flux as technologies continue to collide with one another, forming new ones. A recent example of this is the use of machine learning and instant messaging to create 'chat bots', in essence an 'ask and answer' mechanism using instant messaging but also now being used to deliver search results, book tickets and perform other functions in the social and work environments. Many businesses (and technologies) are acting like balloons under pressure – sometimes the balloon bursts and other times (depending on the location and other forces acting on it) it morphs into a new shape. Data is driving a lot of this change – data needs input and input comes from recording what is happening, which requires sensors. With billions more sensors flooding the market over the next five years in devices like smartwatches, phones, cars and household items (commonly known as the 'Internet of Things'), we are about to see entirely new emerging technologies and economies springing up and altering those that already exist. Some of these new technologies will change industries and others will add to existing economies – but not all will disrupt.

Is it emerging or disruptive?

Both emerging and disruptive technologies can be confusing at times. Emerging technologies are referred to as such because they are not yet fully formed and usually never fully develop a final state of being. Social scientists Rotolo, Hicks and Martin (2014) point this out in their description:

> [an emerging technology is] a radically novel and relatively fast-growing technology characterized by a certain degree of coherence persisting over time

and with the potential to exert a considerable impact on the socio-economic domain(s) which is observed in terms of the composition of actors, institutions and patterns of interactions among those, along with the associated knowledge production processes. Its most prominent impact, however, lies in the future and so in the emergence phase is still somewhat uncertain and ambiguous.

Disruptive technology is a contentious theory with several critics. The main arguments are around when the term should be applied, although as more and more technologies arrive that appear to disrupt, two things become clear – speed and totality are key for true disruption to occur. The speed of the disruption, or rather if the disruption is expected, is key for many; for example, while they are transformative, electric cars have not been truly disruptive because of the way gas companies and other organizations have slowed their progress. In other words, there must be some element of speedy bypassing, upgrading or replacement of the outdated way of doing something. Personally, I feel disruption is disruption whether it happens quickly or slowly but there is merit in the criteria so let's let the nerds have this one.

The second argument surrounds a technology's transformative ability or how much of an old thing a new thing changes and the value this creates. Rather than speed, this, to me, feels like the key criterion for disruptive technology, as Christensen (considered the godfather of the term 'disruptive technology') puts it:

> The technological changes that damage established companies are usually not radically new or difficult from a *technological* point of view. They do, however, have two important characteristics: first, they typically present a different package of performance attributes – ones that, at least at the outset, are not valued by existing customers. Second, the performance attributes that existing customers do value improve at such a rapid rate that the new technology can later invade those established markets. (Bower and Christensen, 1995)

Put another way, disruptive technologies (or at least the ones that iterate in an existing market or company) often discard what people don't like about a business and ramp up what people do like, such as saving money, saving time, being happier, making better choices, being extravagant... the list goes on. Additionally, companies producing the disruptive technology may add in extra desirable elements to further add value to the end user.

A great way of seeing how disruption works and affects established businesses can be seen in a great quote from Tom Goodwin of 'All We Have Is Now':

> Uber, the world's largest taxi company, owns no vehicles. Facebook, the world's most popular media owner, creates no content. Alibaba, the most valuable retailer, has no inventory. And Airbnb, the world's largest accommodation provider, owns no real estate. Something interesting is happening. (Goodwin, 2015)

The perfect case study is, of course, Uber. Except it isn't when you look at it deeply. While it has displaced multiple entrenched taxi services around the world, it did not create anything new; it just changed the rules of the existing system. While Uber is now building on its network and pushing into new areas such as delivery of parcels, food and gifts, these are all areas traditionally owned by other massive companies. The days of Uber not meeting significant resistance may soon be over thanks to various new laws and readings of old laws affecting their business model. New services like Uber may provide more validity for applying the disruptive innovation criteria, but currently Uber is simply mislabelled in the eyes of many academics.

A better example of disruptive innovation is Netflix. Initially a DVD (remember those?) mail-only company, Netflix helped dislodge Blockbuster and flipped the video rental business forever. Netflix identified issues that Blockbuster (a physical product-based company) simply weren't handling well; the service took a lot of time, and availability and choice were an issue. Fast forward a few years and Netflix responded with an online service that offered fast access, affordable service and bandwidth to serve more customers than Blockbuster ever could.

All may not be well in the Netflix camp, as others are now disrupting their business model by creating similar and different models based on the success Netflix continues to see. This highlights the old idiom that seeing and fixing others' problems is easier than seeing and fixing one's own – a valuable insight for many businesses and no doubt for most people reading this book. Looking at someone else's business and 'fixing' it to create something new that enough people want is a core skill in tomorrow's economy. Beyond simply creating new businesses and products, the TBD framework explained later in this book will also help you to identify areas of improvement (or weakness) in your own business before disruptive forces affect the business negatively.

Disruptive technology is not necessarily about killing off old or 'bad' businesses; as we saw earlier, Blockbuster was not a bad business – the company simply failed to move fully with the times and technologies that encroached upon it. Joseph Bower (2002) explains how companies miss this happening:

> When the technology that has the potential for revolutionizing an industry emerges, established companies typically see it as unattractive: it's not something their mainstream customers want, and its projected profit margins aren't sufficient to cover big-company cost structure. As a result, the new technology tends to get ignored for what's currently popular with the best customers. But then another company steps in to bring the innovation to a new market. Once the disruptive technology becomes established there, smaller-scale innovation rapidly raises the technology's performance on attributes that mainstream customers value.

How do disruptive technologies happen?

We've seen a glimpse into how disruptive technologies come to light but an alternative way to answer this question is to look at creativity and how new things get created. There are three basic ways to create something: copy the thing you want, combine multiple things, or transform something to become the thing you want (you can then apply the other techniques multiple times). As Kirby Ferguson puts it in the excellent 'Everything is a Remix' series (Everything is a Remix, 2015):

> Remix. To combine or edit existing materials to produce something new. These techniques – collecting material, combining it, transforming it – are the same ones used at any level of creation. You could say that everything is a remix.

Disruptive technology can get overly complex, with a lot of detail needing to be ploughed through. However, the technology can also be radically reduced to help us understand it and move forward in a smart and effective way. This reductionist philosophy can be applied proactively or reactively and the theme of clarity through simplicity is one carried throughout this book.

So what disruptive technologies are going to be big deals?

Writing about every technology that will 'blow up', while possible, isn't realistic or necessarily helpful for business leaders. Instead, this book will help you do two things: know more about the big areas and discover a framework that will aid you in evaluating any technology and change to be made to make more strategic bets about the future. However, there are five big bets for disruptive technologies that are important for every business owner and department head to understand beyond simple surface knowledge. Some are in their infancy (artificial intelligence, nanotechnology), and others are still on the periphery of mainstream culture (blockchain); some remain misunderstood (3D printing) and some, while highly desired, are not quite ready to take any giant leaps forward (holography).

Why were these technologies chosen?

Some of the technologies are software based, others are materials, and some are virtual. Some names will no doubt be familiar to you and others will not be, but none of them are difficult to understand when you reduce the basic elements down. Each of the technologies has been chosen because of its potential likely impact on business and culture at large. Additionally, the technologies explored offer the biggest potential for businesses when it comes to cost savings, product innovation and future-proofing business functionality.

To aid comprehension and the use of the information, each emerging technology has:

- a clear and simple description;
- a brief discussion of why the technology will impact different businesses;
- a predictive timeline of when they will impact different businesses;
- pros and cons;
- an impact score (out of 10) regarding how big an impact it will have on mainstream culture (or life).

The technologies covered include:

- blockchain (the computation behind Bitcoin) and Bitcoin;
- artificial intelligence (and machine learning);
- holography (including virtual reality and augmented reality);
- 3D printing;
- nanotechnology (including graphene).

Impact scores have been assigned to all technologies based on several factors including likelihood of happening, time to disruption, the potential severity of the impact and the length of disruption (among other elements).

The timeline to impact is a guide for businesses to use as markers to create changes needed to survive and thrive. 'Consumer' relates to when the average consumer will have greater than surface knowledge of the technology and will probably use it – in some form – in their daily lives. 'Enterprise' refers to the other end of the scale, when businesses will be using (or benefiting most from) the technology.

Blockchain and Bitcoin

Often misattributed as Bitcoin itself, the blockchain is like a public ledger of transactions and is the basic element that underpins (or powers) virtual currency technologies like Bitcoin. Often referred to as 'crypto-currencies' or 'decentralized digital currency', Bitcoin can be used in the real world to buy goods but is primarily used for online transactions in both the light and darker areas of the web. While the potential for blockchain technology is far from limited to virtual currencies, it is most famous for driving the idea of virtual currencies and is what we'll focus on here.

The simplest element of a blockchain is a 'block'. Blocks are essentially a permanent record of files that hold data about digital transactions. Each time a block is 'full' it gives way to the next block in the blockchain. No one can alter the contents of a block (part of what makes the technology secure) and every block has a highly complex mathematical problem attached to it. Once these equations and calculations are solved (a process called 'mining') using an individual computer's processing power (or series of networked ones), the miner accrues a Bitcoin he or she must keep in a wallet (think bank account). Every time this is done, a new Bitcoin enters circulation. The harder the mathematical problem, the longer it takes to solve, and so the creation of new Bitcoins is somewhat regulated but not in the way the outside world regulates currency.

The blockchain is stored in networks of distributed nodes across the internet. Each node has a copy of the entire blockchain and as new nodes come and go this secures the chain against issues like poor connectivity, hardware failure or outside forces looking to disrupt the process. In other words, there is no single point of vulnerability, which makes blockchain harder to attack (and less likely to fail) than, say, a centralized banking system owned by one person.

Bitcoin has had a chequered past and, as with a lot of technologies, it continues to see some darkness before it will see mass adoption, notably being used by unsavoury types on the Deep Web (content that is not indexed on the web) to buy and sell drugs, guns and other illegal services. Despite there already being pizza shops in New York that have Bitcoin ATMs and coffee shops that accept Bitcoin as payment, the idea of the blockchain is a very technical one that many people and businesses simply do not want to trouble themselves with yet, although it is interesting to note that such technologies can often be used in stealth. In other words, the user doesn't know they are using the technology. In the case of blockchain technology, this could easily be the case if it was built into a 'super app' (an app that provides multiple, seemingly unrelated, services via a single mobile interface). Sometimes described as services marketplaces, these apps often have hundreds of millions of users and are ubiquitous as users can pay their rent, hail a car, and chat with friends using the same app. Examples of super apps are WeChat (China) or Gojeck (South Asia), Snap (US), Zalo (Vietnam) and Alibaba (China) to name a few. Both Meta and Amazon have also been said to be following super app strategy playbooks, which has caught the eye of EU and US regulators.

QUICK SNAPSHOT: BLOCKCHAIN

What: The underpinning technology of virtual currencies that uses a block to form chains as mathematical problems that are solved by computers across the internet.

Pros:

- *Flexibility*. Due to their digital nature, it is easy to distribute and transfer money or Bitcoins anywhere in the world at any time.

- *Global*. You don't have to worry about crossing borders, rescheduling for bank holidays, or any other limitations one might think will occur when transferring money.

- *Control*. There is no central authority that controls Bitcoin, so you are totally in control.

- *Secure*. Created to ensure personal information is kept private, blockchain technology activity protects against things like identity theft.

- *Lack of fees*. Both a pro and con – while there are no or very low fees, new services can be created to charge extra (e.g. faster processing).

- *Fewer risks for merchants*. Bitcoin transactions cannot be reversed and do not contain personal information, so merchants are protected from fraud.

- *Business in troubled areas.* Due to the blockchain, it could be possible to expand into otherwise dicey areas where it would have been easy to mislead people, because of the way the blockchain is set up. Therefore, Bitcoin could be said to have the potential to transform whole -industries – from finance to retail – as dependence on old money hierarchies drops.

Cons:

- *Level of consumer understanding is still low.* This is a key area of concern for many – while more money for education will likely in effect benefit everyone, no one wants to spend their own budget educating other people's customers.

- *Level of consumer trust is low.* A key concern to users and creators alike, significant resources can be misspent if companies fail to meet the needs of consumers early.

- *Scale.* Growth will likely be slow and hard unless it is adopted by large-scale organizations.

- *Issue resolution.* No system is perfect but since there is no owner, who is at fault if issues arise? Who will sort the problem out? Key considerations for customer service, training and brand reputation management.

- *Volatility.* Blockchain technology, while secure, often means there are limited amounts of coins because of the way it was created. This volatility causes fluctuations and uncertainty and while as time goes on this issue is likely to decrease, businesses will still have to deal with the repercussions.

- *Uniqueness.* Unlike some technologies, Bitcoin and blockchain are hard to grapple with (despite being relatable to currency) and therefore have a higher bar than other technologies like contactless payments and in-app purchases.

Why important? It is not hyperbolic to say that blockchain and the technologies it enables have the potential power to disrupt entire countries. Big banks and corporations have billions invested in maintaining the current state of the financial system but it is when the friction of using Bitcoin and the other technologies reduces that we will really see consumer adoption and larger disruptions occur.

Impact score: Likely = 9. Potential = 10.

(Note: This score is for mainstream life and the average business – naturally (as mentioned above) some businesses will be (or could be) affected more than others.)

Timeline to impact: Consumer = three years; enterprise = two to four years. Mass adoption may be slow due to the level of knowledge and education required, the closeness to something of value (people really avoid messing with money) and finally the protection the current system has from multiple powerful (and interconnected) entries. Look at Chapters 10 and 11 for more information on how blockchain is going to change the next era of the internet and the services we use.

Artificial intelligence (and machine learning)

If we look at film history – *Chappie, Her, I, Robot, A.I. Artificial Intelligence, The Matrix, Transcendence, Blade Runner, Ex Machina* and of course the dreaded HAL from *2001, A Space Odyssey* – things are not going to end well for humans according to Hollywood. To fully understand the realities of AI you need to do one thing and that is erase everything that Hollywood has taught you or everything you have read about AI. The reality, as you'll see below, is much less advanced than we are being led to believe by the news and the movie studios.

At its simplest, AI is any technology (not just robotics) that aims to emulate intelligent human behaviour by appearing to understand complex content, engaging in natural conversation with people, learning and making 'its' own judgements. The applications of such technology are as far-reaching as they can be unnerving, from cars that drive themselves (autonomous vehicles) and speech recognition (customer service bots) to risk detection and consideration. Besides these useful elements, there is also the ability to process much more information – and create useful outcomes from it – than the human brain is currently able to do. The processing element enables systems to run millions, possibly billions, of scenarios and choose the best outcome based on rules we give them. However, it seems it is the sentient nature of such systems that concerns Hollywood producers – rather than being a robot slave, true AI seeks betterment and a focus on awareness beyond simple scenario planning.

Often AI is simply attributed to the loner genius in his basement who stumbles upon sentient intelligence or robots that for some reason develop a glitch in their programming that suddenly gives them human-like awareness. Sadly, the reality is far duller – research is slow, highly technical, siloed and incredibly secretive at the top end of the spectrum as scientists and researchers are often unable to share or learn from other experts because of non-disclosure agreements. Progress is slow and, despite the sensationalist appetite for the technology from some (Hollywood), the reality of a world with sentient robots – barring a breakthrough – is some way into the distance.

However, it would be foolish to ignore AI as being too far in the future to be useful. Meta has successfully piloted 'M', the helpful service within Messenger that helps book tickets and hotel rooms and answers simple questions but with human oversight. In essence, it is not true AI but a learning algorithm that can then predict things based on this data – this is known as 'machine learning' and evolved from attempts to create AI.

The main goal of the AI field remains firmly general intelligence (which is fairly universally accepted as a long-term goal) over some of the subfields like computational intelligence and machine learning that have goals that are easier to achieve.

QUICK SNAPSHOT: ARTIFICIAL INTELLIGENCE

What: The field of science – not limited to robotics – that handles technology aiming to emulate human behaviour.

Pros:

- *Accuracy.* Due to increased processing capabilities, better decisions can be weighed and then chosen.

- *Human limitations.* Creating smart robots does make sense when we think about the human body and its fragility. In particular, the areas of space and underwater exploration have a lot to gain from the use of AI.

- *Freedom.* Intelligent machines can free us from boring jobs and indeed manage the process. While this scenario may scare some, it will equally excite others as new jobs and opportunities emerge.

- *Smart use of time.* Time is something we as a species can never get back. Thus, creating or utilizing tools that can help us do things more efficiently (GPS, predictive text, virtual personal assistants like Apple's Siri) should be the priority to maximize our time here and the impact we have on the planet.

- *Always on.* Robots and AI do not need sleep as humans do, which could lead to significant productivity gains through maximizing the workable hours in a day.

- *Safer.* AI and robots can complete tasks without feelings, eliminating human error due to boredom or tiredness.

Cons:

- *Cost.* AI is incredibly expensive to create and while maintenance is small, updates and changes will be frequent.

- *Ownership.* Machines are programmable units; without consciousness, the robots do as they are programmed to do, so whoever owns the robots could, therefore, be said to choose – do I use them for good or bad? Hacking is also a real concern.

- *Ethics.* As mentioned briefly above, there is a raft of ethical and legal issues to contend with surrounding the giving of life, slavery principles and numerous other 'what if' scenarios. These are large and robust issues that cannot be ignored when thinking about any element of AI. Each has a book's worth of arguments for and against it and while it is always important to explore such things in detail before large decisions are made, this book's aim is to simplify as much as possible, so these will simply be listed here.

- *Data loss.* Due to the significant reliance on big data, the many possible issues faced by AI programs and robots will mean lengthy downtime and costly offline time to restore data and files.

- *Creativity.* Currently, a machine is a machine, but what if that machine is required to think outside of its programming? What about common sense?

- *Emotional intelligence.* Robots have no souls. Can empathy be programmed? This and more questions besides are considerations for when AI is inappropriate. For example, an AI surgeon may technically be a better doctor but would you want it to give you some life-altering news?

- *Degeneration.* What happens if we use our brains less as a species? What impact would this have on future generations?

Why important? AI is the next evolution of computing and potentially the next wave of humanity – imagine what the world could look like if every decision becomes optimized for the best possible outcome.

Impact score: Likely = 10. Potential = 10.

Timeline to impact: Consumer = three years for full public rollout but limited or low-level functionality; enterprise = two to four years for full rollout but limited functionality; +15 years for mid–full AI due to the level of investment required, the level of secrecy and the siloed nature of the field. Additionally, new laws will be required regarding AI along with significant ethical guidelines for its use which will take time to create, agree on and sign into law.

Holography

Holography is simply the study or production of holograms. While often confused with tomography, Pepper's Ghost Illusion and volumetric displays, holograms are created using a technique that reads the light from an object and then presents it in a way that appears three-dimensional. It's important to note that the term 'hologram' can refer to both the encoded material and the resulting image – whether static or moving. While it is likely you've seen stickers, security features and images that appear holographic, it is unlikely you've had a jaunt in the Holodeck from Star Trek or a conversation with Princess Leia from Star Wars (although the ability to do both things is tantalizingly close, in laboratories at least).

Recently, holograms have come back into favour within multiple industries, especially music, although these utilize a technique known as Pepper's Ghost, based on an old trick that fools participants into thinking they see a ghost appear out of nowhere by using angled mirrors. This technique has been used recently with Abba, Michael Jackson, Tupac, Mariah Carey and Elvis Presley and although this is not a 3D hologram, these examples are an interesting halfway step to what will one day likely be commonplace.

When you sit back and think about holograms, there are several issues. Beyond simple technical issues, there are multiple problems surrounding the legal use of likeness, distribution rights, fraud and the high costs involved in large-scale images without 3D glasses. In labs in Japan, the technology is being pushed even further with the use of haptic technologies (air pulses, ultrasound) that further extend our experience with holograms as they are not yet physical experiences you can interact with.

Most 3D holograms to date have been relatively primitive when it comes to design although this will change rapidly as more people gain access to the technology and begin experimenting with it.

QUICK SNAPSHOT: HOLOGRAPHY

What: Holograms and volumetric displays.

Pros:

- *Multiple applications.* Holography is a burgeoning field with lots of areas that can benefit from the insights and technology, including terrain modelling, scientific visualization, medical visualization and architectural modelling.

Cons:

- *Expensive*. Currently, the technology required to create holograms and 3D imagery is limited and expensive due to the technology and man hours required.

- *Limited need*. Currently, while there are benefits to using holograms, they are by no means essential. Beyond consumer desire to have Princess Leia-esque messaging, the use cases often do not match the cost or effort required.

- *Time-consuming*. Creating holographic images takes a lot of time due to the multiple processes required and the planning involved.

- *Technical issues*. Due to the physics involved, holograms will never work well in sunlight.

- *Lasers*. Current technology uses lasers to create holograms which, if viewed from incorrect angles, have the potential to damage retinas permanently.

- *Ethics and rights*. Currently, using people's likeness is covered by multiple laws and regulations – holography (especially) after death occurs causes another legal headache for creators and users of the technology.

Why important? Holography, because of its Hollywood portrayals, has been regarded as the holy grail of technology. Whether recorded or live, holography is one technology everyone wants but perhaps isn't sure what to do with it once they have it. True holograms (i.e. without the aid of a visor or another screen) are a technically challenging arena that can be extremely helpful for things like crime scene visualization or used for entertainment purposes. Both are valid but do not have equal value for society.

Impact score: Likely = 5. Potential = 8.

Timeline to impact: Consumer = +12 years; enterprise = +10 years. The long timelines are due to the technical nature of holography and the current state of the industry. While holograms do exist, current technology limits mean they are rarely seen outside of a laboratory. Beyond this, their practical applications, while not limited by any means, are equally not crucial to our existence – for example, dentists have X-rays and other imaging technologies to which a hologram wouldn't necessarily add more definition or understanding. Beyond these issues, the costs are simply too high currently for a speedy uptake in either the consumer or enterprise sector.

3D printing

3D printing is the process of using a digital file to create a three-dimensional solid. It is often referred to as 'additive manufacturing' as it involves adding layer after layer of material to create the final product. Currently, most 3D print manufacturing is at a somewhat low level, in most cases used for replacement equipment parts or prototyping new designs for trainers.

An easy way to understand the complex process is to imagine a piece of string being folded back and forth focusing on different areas sometimes more than others to create a model of Michelangelo's 'David' from toe to head. The process itself is pretty simple. First, take the design of the object you want to print after it has been digitally rendered by powerful computer programs or by scanning an object with a 3D scanner. The next step is to print the object using a 3D printer. There are several types of printer that use multiple processes but mainly vary the way each of the layers is added. For example, some methods use a melting process to create the 'thread' to create the layers (the most common method) but others use lasers to harden a pool of material instead. The materials that can be used in 3D printing include glass, nylon, wax, silver, titanium, steel, plastic, epoxy resins, wax, photopolymers and polycarbonate, with more announced from universities and laboratories around the world on a monthly basis.

Beyond simple trinkets and replacement parts, 3D printing gets incredibly interesting and cutting, nay bleeding, edge, with a large number of fields experimenting with new materials and construction techniques:

- graphene – stronger than steel, graphene is giving a massive headache to the metal industry because of its lightness and strength;
- quick-drying cement is enabling construction firms to 3D print homes and emergency shelters in disaster-stricken areas, enabling quicker recovery for communities;
- human cells are being used to create human organs and body parts using 3D printing techniques – a field called bio-printing;
- food has also been printed – fruit, pancakes, pizza, ice-cream, burgers, waffles and chocolate have all been piped into existence;
- drugs are one of the most recent items added to the list of things that can be 3D printed, with vitamins and epilepsy medication already in circulation.

QUICK SNAPSHOT: 3D PRINTING

What: The process of creating a 3D physical object from digital blueprints that exist in databases and can be downloaded (if the author gives permission) by anyone, anywhere and anytime.

Pros:

- *Versatility.* As confirmed by the list of materials above and the ones being tested, 3D printing is without doubt one of the most versatile disruptive technologies out there.

- *Intricate.* 3D printing has already achieved standards on a par with traditional manufacturing methods in medical fields and has created prosthetics devices and dental accessories.

- *Prototyping.* The ability to rapidly create examples to be tested before mass production lowers costs and saves time in getting products to market.

- *Green.* Most components utilized by 3D printing offer the same (or superior) safety and stability capabilities as traditionally manufactured components but at a fraction of the weight. Now push this idea up a few levels when using metal in aircraft. Consider the ability to save money and the environment due to requiring less fuel and you start to see why 3D printing is the disruptive force it is.

- *Logistics.* If there are 3D printers in Los Angeles and New York you don't need to transport goods between them; simply print one out where it is needed, again saving money and time.

- *Overproduction.* 3D printing is on-demand printing at its core, meaning that you print what you need, saving resources, time and waste. It also means you can print small batches when needed rather than shelving replacement parts in massive (and expensive) warehouses for many years.

- *Customization.* With free designs, low-level entry and increased access to high-level specifications and information, the ability to alter, remix and reimagine increases exponentially. The ability to redesign the latest lamp shape to their own specification is now within the grasp of every consumer.

Cons:

- *Mechanization.* The undertaking of more work by robots has massive implications for the labour industries. 3D printing enables the average consumer to create physical objects that can be used for a wide variety of purposes without the need for specialists, shops, delivery people and many more besides. Beyond this, 3D printing could also bring a fair amount of these jobs home and encourage increasing worker skill levels.

- *Copyright.* Intellectual property and copyright are massive issues in the area of 3D printing, and the law is being slow to catch up. How do designers and manufacturers maintain the value of their goods if a digital file cannot be secured? If a free design is modified and becomes a roaring success, does the original creator see any royalties?

- *New laws need to be created.* Guns have already been 3D printed and have quickly been banned in several states. With all new technologies, making sure people are safe is paramount, but often the law has to run to catch up to the potential (and threats) offered by new technologies.

- *Quality.* While some materials offer superior qualities, not all designs do. Currently, 3D printing designs are shared freely and openly – a great democratizing benefit of the technology – but often quality is simply lacking in the designs.

- *Liability.* What happens when something breaks? Who is liable if anyone can be a producer or manufacturer?

- *Speed.* The big issue with 3D printing is that it is not a quick process, with larger items taking double-digit hours to complete before the additional work – which most items require – can begin. Mass production is still a way off due to the cost of printers – especially the larger units.

- *Cost.* 3D printing is still expensive, and while costs are coming down (both of materials and machinery) it will be a significant number of years before the costs involved will be comparable to traditional manufacturing processes.

- *Control.* If items – especially drugs – can be 3D printed, what happens to the oversight and regulation? What happens to Customs departments when items are no longer transported across borders and oceans?

- *Shipping.* While many analysts predict lower transport costs because of 3D printing in the long term, many predict a rise in closer-to-home transport costs due to items requiring transport from printing hubs or stores.

Why important? The worldwide 3D printing industry is expected to grow to $83.90 billion by 2029 according to data from Fortune Business Insights. Due to falling costs, 3D printing technology has the capability to transform almost every major industry (think everything but banking) and change the way we live, work, and play in the future. The most likely outcome is that 3D printing will take its place alongside traditional production technologies, rather than replace them.

Impact score: Likely = 10. Potential = 10.

3D printing is one of those technologies that seems harmless enough until you begin to think about the ramifications for big businesses of people printing replacement parts and making things last longer, let alone what this sort of technology does for developing economies based around manufacturing. 3D printing is unlike all other technologies previous generations have ever witnessed due to the lack of hierarchy required and distributional elements associated with the field.

Timeline to impact: Consumer = +four years; enterprise = two to five years. 3D printing is already available in a limited format for both consumers and enterprise, but the technology is not mass market yet because of the costs involved, time of production and current materials available. Therefore, the disruption is truly set to come once this peak has been reached.

Nanotechnology

Nanotechnology is all about the small – the ridiculously small in fact. Often referred to simply as 'nanotech', the field derives its name from its scale and relates to the changing of matter at the atomic and molecular level to create new properties and applications.

Fields that are directly working with nanotechnology already include:

- aerospace (new materials, batteries, lighter materials);
- food (better preservatives);
- consumer electronics (scratch-resistant screens);
- energy (cheaper and cleaner sources);
- medicine (faster absorption).

The scale at which nanotech works is staggering. A pin-head (1 mm diameter) could have a million nanometers laid end-to-end across it, a sheet of paper is about 100,000 nanometers thick and a single red blood cell could have 2,500 nanometers laid across it. The desired outcomes and properties that result from breaking down and recombining different molecular and atomic elements are – as you might expect – strength (new metals), speed (electric conductivity) and weight (graphene – a super-light new metal) among others. Following these new properties is a raft of applications that are as desirable to the military as they are to scientific fields and the business world. Nanotech is already in use in numerous commercial products and processes when products require materials to be lightweight yet strong or with specific properties – think sunscreen, sporting goods and boat hulls for common applications.

Once the realm of science fiction, nanotechnology is about to revolution-ize the medical field in particular. The ability to put sensors and diagnostic tools inside the body that send back data to the outside sounds a long way in the future but in fact Phillips recently introduced 'VitalSense', a pill-sized device that continuously monitors a person's core body and skin tempera-tures without touching the patient. In essence, it is an indigestible device rather than a wearable one. Dentists, opticians and pharmacists are all in some way, shape or form also using nanotechnology, be it to improve absorption, create scratch-resistant coatings or to promote bone growth from surrounding cells.

There is a global arms race (of sorts) happening when it comes to nano research. The United States currently spends £13.2 billion annually on nanotechnology, outspending China around 3×, but the gap is closing. Based on Allied Market Research data, the global nanotechnology market should increase from $5.2 billion in 2021 to $23.6 billion by 2026.

QUICK SNAPSHOT: NANOTECHNOLOGY
Pros:

- *Better attributes.* While stronger, lighter and cheaper doesn't always mean better, it is likely that the flexibility of the nanotechnology properties will improve a wide variety of fields.

- *Impact.* If things last longer, we should be able to buy fewer of them.

- *Recycling.* Advanced nanotechnology is currently being researched that will clean up landfills, in essence 'eating' (or molecularly destroying) rubbish – taking recycling to a whole new level.

- *Health.* Internal medicine will be massively affected by nanotechnology. Fields such as nutrition will also likely see benefits, with 'smart' foods that could potentially fight disease or stop ageing. In essence, the nano-technology is programmed to seek and destroy various ailments.

- *Preemptive.* If nanobots can 'live' inside us, the ability to monitor and combat sickness early increases exponentially.

Cons:

- *Weaponization.* Nanotechnology has the potential to be lethal if turned inwards towards the human body. Issues surrounding the invisibility and programmability of the technology have many people concerned in this regard.

- *Human costs.* With nanorobots and self-replicating technology on the horizon, human jobs are at risk as robots can work harder and longer under poor conditions.

- *Health issues.* Nanoparticles used in paint have been found to cause a severe lung disease in paint factory workers (Smith, 2009).

- *Environmental effects.* Despite being microscopic, the potential contribution to environmental destruction cannot be understated – especially if self-replication of nanobots (where one nanobot creates other nanobots) becomes a reality.

- *Economics.* Nanotechnology is not cheap and the research is not universally available, meaning intellectual property and skills become a rich area for disruption at a macro and market level.

- *Control.* Any country or business that forges ahead in this area will potentially have a significant advantage over others.

Why important? Despite the highly complex processes and costs involved when creating nanoproducts, the future for nanotechnology is bright. The possibilities when it comes to health and the human body are staggering and continue to drive massive amounts of research into prolonging and improving life. Academics predict the technology will evolve into new areas, including being capable of self-replicating (collecting other particles it needs to create more 'nanites') – a theme already picked up on by Hollywood.

Impact score: Likely = 7.5. Potential = 10. The potential for nanotechnologies is impressive because of the versatility at the heart of the technology and the power it offers to business and humankind. While the technology has steep learning curves, costs and implications, it has the potential to revolutionize multiple industries.

Timeline to impact: Consumer = five to ten years for advanced property consumer nanotechnologies; enterprise = five years. As with 3D printing, nanotechnology is available but has serious cost implications attached that are holding back larger breakthroughs and application in consumer settings. Disruption could occur at any time so make sure to keep an eye on research that is happening in your field.

Conclusion

Each of these technologies represents some of the biggest opportunities and major changes the world will see in the next few decades but none is without issue. Now that we have looked at each of the technologies to understand what they are, what they could become and when disruption will happen, we need to look at the larger landscape. These technologies are constantly changing and colliding into one another. In the next chapter, we will look at the brutal truths behind these technologies, why they are often misunderstood, the barriers to putting them into practice in corporate environments and why bad technology choices undermine customer relationships.

Beyond understanding how technology impacts the customer relationship, Chapter 2 discusses the need for a flexible approach and understanding of these and future technologies. This framework has become known as TBD and isn't set in stone; it is simply a pin to help navigate a route forward. The route will likely change but the key is to have a flexible system to enable you to move forward at all stages.

02

Disruptive and emerging technology: the brutal truth

What's the problem here?

The issue at play is the customer experience and perception of your brand. Whether it is a product that doesn't work, doesn't work for long or increases the time to the desired outcome (leaving quickly, paying securely), bad technology is causing a lot of headaches for companies and consumers alike. Most of us have felt the frustration of not being able to pay via contactless means and while this is dubbed a 'first-world problem' it is a big one for companies looking to maintain and grow loyalty in a world where people's attention is at an all-time low. Ultimately, poor technology – whether you create it or use it – affects a business's bottom line. From websites to point-of-sale, everything a consumer touches has an effect on their perception of your brand, its competency and whether they should trust you (now and in the future). Investment in emerging and disruptive technologies therefore becomes a critical area to consider in all budgets if your company wants to survive into the 22nd century.

The brutal truth about emerging and disruptive technologies is that they are here to stay; in some cases they will increase in speed and impact and beyond this there are more unknowns than there are knowns. In essence, we don't even know what we don't know.

Donald Rumsfeld summed this up perfectly when he talked about weapons of mass destruction (US Department of Defense, 2002):

Reports that say that something hasn't happened are always interesting to me, because as we know, there are known knowns; there are things we know we know. We also know there are known unknowns; that is to say we know there are some things we do not know. But there are also unknown unknowns – the ones we don't know we don't know.

The fact that there is so much uncertainty associated with technology was a key driver for writing this book; once you are comfortable with this element, great things can (and do) happen. This does not mean you have to take huge risks (sometimes you will) but having a clear understanding of what you know and what you don't enables you to at least make judgements and keep moving forward.

As we've seen in the Introduction and Chapter 1, emerging technologies differ greatly from truly disruptive technologies. They will continue to be mixed up and misattributed but hopefully you can now see and benefit from understanding the difference between the two. As you move forward through the remaining chapters, technology as a whole will be spoken about with a focus on the disruptive elements.

Disruptive technology is misunderstood – or worse, ignored – by many because it is often too technical or takes time to understand. An example of a technology that usually falls foul of these issues is something like 3D printing. The results can be seen but the issues are behind the scenes and therefore difficult to work through. Working with multiple senior executives at large and small companies across a wide spectrum of clients has taught me one thing: understand the person before the technology – a point explored in more detail later on in the Behaviour section of TBD.

Beyond simply putting off what is difficult or unknown, many executives simply don't see disruptive and emerging technologies as part of their job, but rather as something special that should be kept for a once-a-year type of extravaganza.

While this approach can be galvanizing and inspiring, it is often what can be termed 'firework activity' – it burns brightly and impresses but is quickly forgotten and must truly stand out in order to be remembered and not block a person or company from moving forward.

None of these blocks are insurmountable. A few of them are difficult and require discipline – the latter issue ('it's not my job') needs careful management as it is often a sign of larger issues or an employee who is unhappy or lacking the skills to be open to change. Use this opportunity to move everyone forward.

It doesn't have to cost the earth

The other misconception about emerging technologies regards the high price and cost of continually staying ahead of the competition. While this point can be true for some companies because of the technologies they will be interested in, most can, with the right goals, create a smart innovation and emerging technology programme for very little time and money. Sometimes understanding the technologies available and waiting for others to educate the masses is the correct strategy; other times you will need to lead. Understanding when to and when not to invest money is crucial and we will cover this more in Chapter 4 when we look at the Decision Matrix.

Small is beautiful

A common response when a new technology strategy is questioned by senior executives is, 'We don't need to be first.' This is a fair answer, as a lot of the bigger companies have dedicated budgets and reasons to lead. However, this scenario is likely to change due to the agile nature of start-ups and the way new technologies can be rolled out faster and faster with established networks already in place. In a lot of ways, many small businesses are leading the way for industries or other businesses because of their size, not despite it. Financial technologies (often referred to as 'Fin-Tech') like Stripe, Square and iZettle that focused on transforming the way small businesses took payments from customers across the world are a great example of this sea change. These three companies are now much bigger than their original products which were designed to make digital payments easier either by creating a simple but technologically advanced point-of-sale solution for businesses of all sizes (Square) or reducing the friction of selling through social platforms like Twitter and Meta (as Stripe have done).

In this case, the creation of an agile steering group with a key lead figure is crucial to short- and long-term success.

Commitment is key to success

The final area where many companies let themselves down when thinking about emerging and disruptive technologies can be summed up in one word – commitment. Many companies I have worked with did not have the drive in the beginning but had clear needs, whether they were financial, retail development or issues with diversifying product line. Top executives simply did not view emerging technology as part of their ongoing strategy and focused on short-term objectives that were in front of their face rather than forces arriving on the periphery. This is an easy and all-too-common mistake for any business in this fast-moving world that works in the short term but ultimately leaves you in the same or a worse place year after year. The job here is simple. Ask yourself, 'What is the risk of doing nothing?' and whether there is real commitment to completing the task at all levels. A great way of thinking about this if your company is in a state of inertia or not moving in any direction is to ask yourself, 'Are we doing good work or are we just comfortable?' When I go in to see a business for the first time, the conversation often falls into one of those two areas: as a business coach once said to me, 'Your job is to afflict the comforted and comfort the afflicted.'

It starts with you

Often individual workers – that means you too – don't see disruptive and emerging technologies (including simply keeping up with existing technological changes) as an integral part of their job description and therefore deprioritize research (or simply taking time to think) compared to more 'business as usual' tasks. Usually there are two types of push backs when it comes to emerging technology. Some simply push back on the time it is perceived to take, saying, 'I have no time to do the job I am asked to do right now, let alone "this stuff"', while others simply don't believe it is their job to do this sort of thing. Neither response means those people are bad employees or couldn't do the job, but some adjustments are required so your programme is given the best chance for success. Later on in the book there will be a checklist of things you can do personally to keep you and your colleagues updated on disruptive technologies that don't cost the earth or suck up too much time.

The answer to the issues of both time and money is simple: re-evaluate the available resources, namely time and focus. These audits enable the individual to find the time to identify areas of weakness and provide senior executives with evidence if additional resources are required. The exercises are also great ways to identify areas where time is being abused or lost to frivolous activities at work (and beyond). Below are two methods of doing this:

METHOD 1

As Laura Vanderkam, author of *168 Hours: You have more time than you think* (2011), says:

Recognize that time is a blank slate. The next 168 hours will be filled with something, but what they are filled with is largely up to you. Rather than say 'I don't have time', say 'It's not a priority'. Think about every hour of your week as a choice. Granted, there may be horrible consequences to making different choices, but there may not be, too. Dream big. (Vanderkam, 2015)

Vanderkam goes on to discuss thinking about what you would like to do with your time in terms of personal goals (travel, professional, family and so on). Vanderkam also notes the importance of focusing on what you would like to fill your time with and creating a huge list, separating it out into work, family and life columns. Some items will come together, some will stand out – the goal is to write them down and keep coming back to the list to keep you accountable. A good tip is to keep the list somewhere you will see it often – next to a mirror, on a card in your wallet or by the coffee machine if you make coffee every morning. Make looking at the list part of your ritual and you will start to make things on the list happen by seeing connections and opportunities you may have otherwise missed.

METHOD 2

Another option is to go the old-fashioned route and use a paper-based accountability system:

Step 1: Print out a 12-month blank calendar (or buy one).

Step 2: Decide how you will mark your calendar – some simply write in what to do when, others decide to colour-code tasks (work, meeting, new business, research and so on) to see a more visual representation of time.

Step 3: Set a reminder in your calendar, smartphone, smartwatch or other device to always fill in your calendar for at least a week but ideally two or more weeks.

Step 4: Set aside time to evaluate the data you have collected. Out of 168 hours per week, how many have you spent at the office? How many were spent working? How many were spent on work that really mattered? How many were spent in meetings? How many were spent in meetings you didn't need to be in? How many were spent with your family? How many were spent doing what makes you happy? However you decide to measure yourself.

Step 5: Figure out the changes that need to be made. This is the tough bit – as Vanderkam surmises above, 'What is a priority? What are my priorities at work/home? What's my strategy to stop doing X and do more of Y at work/home?'

> **Step 6**: Implement. This could mean having some conversations with your boss, your loved ones or simply holding yourself accountable. Sometimes this can be a simple phone notification or a written reminder in a diary or ritual you do at the beginning of every meeting. The system works if you commit to it and reinforce your good behaviours. Stick with it!

Often when employees are asked to do more, conflict and resentment can occur because of the way things are 'handed down' or the way the news is presented. Here are some tried and tested ways to make sure the news is received positively.

Do:

- Lead by example. It sounds obvious but usually interest trails off after the initial announcement of the activity. It is essential that you demonstrate willingness to be involved, lead by example and encourage others to do the same. Every corporation and business is different but the key to not looking like someone who hands things down is to get your hands dirty regularly. Instead of simply doing something towards the goal, ask people whether they think it should be done or how it should be handled. Involving more people on a regular basis means you are more likely to succeed.

- Pick the right tool for the right job. While tools like meetings, email and IM may seem obvious, they may also be part of the problem. Being able to take time out to discuss issues and options is important; make sure the human touch is not lost when talking about technology – it often is, which is why I built it into the HERE/FORTH framework (behaviour). Slack (the 'email killer' that replaces email with group working and instant messenger) has become a great halfway house – quickly enabling groups to make decisions but also decide when to 'take things offline'.

- Create a steering group. This is simply a group of the key people needed in order to effect change in a department or larger entity. Pick a lead or several leads so there is built-in redundancy in case people leave or get sick. In larger organizations, steering groups are key to enabling change and we'll talk more about this in later chapters.

- Use words and phrases like 'us', 'we', 'as a team', 'all of us', 'so we', 'when we', 'when we get this right...', 'this is key to our...'.

- Reward good behaviour and do it publicly. A weekly 'winner' isn't recommended or likely to be necessary but an unexpected presentation of a small token at the end of a lunch and learn often goes a long way to motivating others. Options other than the standard gift card could be a subscription box from sites like:

  - birchbox.com (mainly skews female);
  - notanotherbill.com (great variety);
  - bespokepost.com (mainly skews male);
  - escapemonthly.com (food, gifts and items from a specific destination).

- Identify the alpha influencers within your organization. These are the people others naturally rally around, follow and do things to support or show loyalty to. It is not easy to influence them but with their support, new initiatives have a greater chance of acceptance.

- Make it part of employee reviews. Sometimes this can be a double-edged sword but – depending on the type of culture your business has – it can be transformative. In essence, it gives the employee permission to develop, demonstrate and focus on improving and highlighting new skills as part of their job.

Don't:

- Get angry – be understanding. All change is difficult and job descriptions are easy to cling on to. Foster a culture of change acceptance by introducing it early and making job descriptions more fluid in nature but with clear focuses.

- Forget to celebrate small victories. Change can come fast and big changes can happen but there will most likely be lots of small changes. Take time to draw attention to these mini-milestones and make sure people understand why they are important and the part they have played in them. Some examples of mini-milestones to celebrate could include X amount of content shared, new products launched, some sort of internal awareness recognition, days saved, knowledge demonstration, or use of new business pitches.

- Just leave it up to the employee. For the maximum chance of success, you should offer them options or some preferred routes. In the case of simple knowledge increase, this could be setting up a Flipboard magazine, creating a Twitter list of influencers for the team to use, or subscribing to *Wired* magazine and presenting interesting articles in a scanned PDF on a monthly basis. Empowering individuals to increase the knowledge of others – and acting like an editor – is a skill a lot of people miss and really benefit from.

- Assume people know why you exist. I often will ask lower-level employees what they think the company's motto or beliefs are; this helps to not only gauge the level of engagement, understanding and commitment in the company, but also to see how well the internal training is working. Ask a few people (or have someone else do it subtly) and you'll have a great benchmark to begin from to ensure everyone understands the importance of this activity to the overall objective for the company.

What is bad technology?

There are numerous essays on the subject of what makes technology good or bad. Yet many fail to address the core issue that surrounds technology today, namely our use or application of it. Technology is not inherently bad. It is the way that it is used or applied that leads to bad repercussions such as obesity (less physical movement), job loss (robots are replacing workers) and privacy invasions (cell phones) to name but three. Not all technology is created equal, however. Some is inherently more likely to cause issues or upset various parties. This is often particularly the case with disruptive technologies because they challenge heritage and protected systems, seeking by their very nature to break controlled arenas.

Bad technology could be said to be specific elements that are baked into a product that have negative outcomes such as:

- *Planned obsolescence.* Essentially this means a product is designed to break or need replacing. This practice is designed to create long-term sales and reduce the time it takes to generate another sale or consideration cycle.

- *Copyright infringement.* Apple, Microsoft and any other technology company out there lose millions of dollars every year due to knock-off or copyright-infringing technology. A related issue, lying beyond just simple copyright infringement, is that poorly sourced and created parts are often used at the beginning of the process to keep costs down, which means the technology can often be faulty and distributed widely. This then impacts on the original brand as the consumer is often unaware of the forgery and will associate the poor experience with the originating brand.

- *Environmental issues.* Technology requires resources in order for it to be created, distributed and sold, which are all then multiplied if new versions are constantly needed. A lot of the materials used are finite (fossil fuels, paper) and therefore it is in all our interests to buy products that are from sustainable resources. However, these are often more costly, leading to the consumer more frequently choosing the cheaper product over the sustainable one.

Other issues that are often attributed to technology but could be said to have a primarily human cause include:

- lack of ergonomic design;
- health issues (poor sleep, repetitive strain injury, migraines);
- shortened attention spans;
- lack of social interaction;
- a warped sense of reality;
- poor social skills and etiquette.

There are a whole host of others. Technology itself is not to blame here, but our use (or rather, overuse) can be. Boundaries need to be set and good behaviour (skilled design, for example) needs to be rewarded and championed.

However, good technology can still go bad and a great example of this is encryption. While encryption to keep online transaction data secure and unhackable is a perfectly good goal for any technology, it is also a magnet for people also looking not to be tracked. A prime example of a technology that was misused in such a way is Tor. Tor is free software that enables people to conceal their online location. Initially launched to stop things like ad-tracking and snooping from undesirable parties (such as cybercriminals), it has been attributed as a major force behind the rise of 'the Deep Web' (the part of the web used to buy and sell drugs and other unsavoury items and services).

Another example is Google Earth. The tool that Google created to enable people to navigate and explore the world around them (or view locations they would like to visit) has helped businesses and economies grow because of its functionality but it has also allowed unsavoury types (such as terrorists) to have access to detailed imagery of areas that would otherwise be difficult to view or analyse. For this reason, Google has implemented a system of blurring out 'sensitive areas', a status that it continually evaluates. Google cannot control how parties use the information so their liability is removed.

So are we doomed to a world with good technologies but bad people?

Most likely the answer to this statement is 'yes', but there is hope. Paraphrasing Rumsfeld's earlier comments, the future offers us multiple knowns and multiple unknowns, but the first part is interesting when it comes to good technologies and bad people. Innovations in the computing space – with a particular nod to the artificial intelligence and machine learning spaces – are likely to lead to 'self-healing' networks that can detect intrusions by hackers, schedule repairs and shut out the offending parties. While this can be both a positive (less down time) and a negative (systems can be difficult to take offline) the outcomes of technology remain firmly with humans for the considerable future. Often, it is how humans deal with technology that is the issue. A good example is machine learning; as machine learning continues to be mixed up with artificial intelligence it is important to remember that *Terminator*'s Skynet scenario (computers attempting to kill humanity) is a possibility but a distant one and one not based in science, for now at least.

TBD is the solution

TBD stands for Technology, Behaviour and Data – the three pillars of a framework I created more than a decade ago and still use with clients to this day. Increasingly, a fourth pillar is coming into the framework (Design).

- **T:** Technology is key because it is crucial to the future of society, business and indeed humanity as we race towards increasing possibilities. All problems have a technological aspect, whether it is high tech (computers, mobiles) or low tech (paper, space); what is important is how we identify, qualify and assess the issues surrounding each element of a problem to make sure they are solved in a sensible way.

- **B:** Behaviour remains a constant in life. While some technologies never come across a human being, most are rarely devoid of behavioural elements that impact the issues and outcomes. Having taken a degree in psychology, the subject has been and always will be close to my heart because every problem has a cause that impacts people directly or indirectly.

- **D:** Data is the element that is often hardest because some of the decisions made or areas to be explored – especially when dealing with disruptive and emerging technologies – lack the necessary data in order for firm decisions to be made. Data is crucial in making smart decisions or even an educated guess. Some data is imperative while other data is a 'nice to have'.

Taking each section individually enables a decision, new technology or issue to be broken down into manageable chunks and scored according to the impact it has on the business. With an inherent focus on the future, risk and potential gains, TBD is a flexible framework. In the coming chapters, it will be broken down before being brought together again in a simplified canvas for you to use yourself when required.

03

The forecasting fallacy

Disruption and disruptive technology forecasting is a multi-billion-dollar industry. Think tanks, trend companies, agencies, research firms, data scientists and of course consultants are all vying for the attention and money of top businesses and brains. Forecasting and prediction are important; they can save companies millions of pounds and make decisions easier. The problem with the array of companies vying for the money is the multitude of approaches available – they are not all created equal and few are flexible enough for today's challenging environments and economies. Often this lack of flexibility means the results can be skewed, unrealistic, or worse, flat out wrong. While this is partly the name of the game, getting predictions right can also save lives and solve massive problems, so making prediction and forecasting better every time should be everyone's priority. Beyond this, prediction and forecasting give the everyday business and individual the ability to make better decisions and possibly stave off invaders, so close attention should be paid if long-term success is desired. Ultimately, the core goal here is change for the better and this chapter discusses how to understand and apply change.

Forecasting is hard but you can make it easier

As mentioned in the Introduction, Philip Tetlock is the grandfather of forecasting or rather 'superforecasting'. His 20-year study showed that even the best experts have off days (quite a lot of them in fact) but there are things you can do to improve their prediction skills. In their most recent book, *Superforecasting* (2016), Tetlock and Dan Gardner detail an experiment where thousands participated in an Intelligence Advanced Research Projects Activity (IARPA) tournament that focused on real-time updates to forecasts.

Using a Brier score (a score that measures the accuracy of chance-based predictions) the participants were ranked, with the top scorers dubbed 'super-forecasters'. Overall, the average Brier score was 0.25; superforecasters on the other hand scored higher, with 0.37 – outperforming (by 30 per cent) the intelligence community analysts who had access to secret data.

Superforecasters mentioned in the research (among others) include Sanford Sillman (an atmospheric scientist), Doug Lorch (a retired computer programmer) and Bill Flack (a retired US Department of Agriculture worker). The plot gets thicker when you factor in that other forecasters were plumbers or ballroom dancers – areas not normally associated with high-level analysis or cognitive function requirements. You could argue these individuals were simply made a certain way, that their brains were wired differently, but the reality is much simpler. All these people are smart, not geniuses (think top 20 per cent rather than 1 per cent), but they have demonstrated multiple times that they are better than people specifically paid to do the job of geopolitical analysis. These people could have been you or me; they were not making lucky guesses but nor had they been trained in specific techniques or had spectacular educations. Therefore, it is sensible to ask, what makes these guys able to do what they do? What can we learn from them?

A large chunk of Tetlock's findings points to a few areas of focus (many of which are baked into the TBD framework which was created many years ago). As with the TBD framework, superforecasters focus on simplicity, or to put it another way, they seek to switch complex questions for easier ones. For example, one could change 'Will the e-book version of this book be bought by eight or more countries?' to 'Will this book be read outside of the UK?'

Tetlock also found that superforecasters were more likely to be able to assimilate lots of data but not huge quantities. Instead they were able to revisit old assumptions when new data was found and adjust accordingly – often with minor adjustments rather than big swings. The largest area of Tetlock's research (and one the TBD framework is also in agreement with) is the 'growth mindset' or a mix of determination, self-reflection and a desire to push through mistakes and learn from them. In other words, being right or wrong is fine but improving is always the goal.

Understanding forecasting is important for businesses because it reduces potential risk, enables the careful spending of money (and resources) but also enables a greater level of planning to occur. However, Tetlock states in *Superforecasting*, 'Beliefs are hypotheses to be tested, not treasures to be guarded.' This statement is a huge area for businesses to consider when forecasting, since most plan for a specific time period rather than using a continuum approach. It is easy for businesses to justify a shorter outlook strategy in uncertain times due to a lack of openness to being wrong (often with good reason). However, this inflexibility leads to poor decisions being made because they usually involve reacting to situations instead of creating more positive ones – either in the long or short term. This is a key reason why TBD was made to be flexible; the framework enables businesses to make ongoing corrections and decisions based on information happening that is either in front of them, just out of reach or further in the distance.

What matters can't be forecast and what can be forecast doesn't matter

Isaiah Berlin – a Russian philosopher (and superforecaster, according to Tetlock) – believes there are two types of prediction specialists: foxes and hedgehogs. Before we delve into more specific characteristics, take this test to determine which you are – go swiftly but don't rush (see also Finney, 2006).

How to determine if you are a hedgehog or a fox

The idea is to agree or disagree with the statements below. Weightings have been converted to a point value (Tetlock rescaled his weightings which were determined by applying another statistical method). Depending on which you choose, you either add or subtract that number. If you agree, give yourself that many points but if you disagree, give yourself the negative points. (Note: Some are already negative points so remember subtracting a negative means adding.)

Your final number should be between –54 and 54. Negative scores indicate you are more of a hedgehog while any positive score means you are more fox-like. As mentioned above (and because of the range) the farther you are from zero, the heavier you lean towards the different thinking style.

1   Isaiah Berlin classified intellectuals as hedgehogs or foxes. The hedgehog knows one big thing and tries to explain as much as possible within that conceptual framework, whereas the fox knows many small things and is content to improvise explanations on a case-by-case basis. I would put myself towards the fox end of this scale.

If you agree +7 points.
If you disagree –7 points.

2 Professors are usually at greater risk of overestimating how multifaceted the world is than they are of underestimating how complex it is.

If you agree −3 points.
If you disagree +3 points.

3 We are closer than many think to achieving explanations of penny-pinching politics.

If you agree −5 points.
If you disagree +5 points.

4 Politics is more cloudlike (unpredictable) than clocklike (perfectly predictable with the right knowledge).

If you agree +4 points.
If you disagree −4 points.

5 The more common error in decision making is to abandon decent ideas too quickly, rather than sticking with poor ideas too long.

If you agree −5 points.
If you disagree +5 points.

6 Having clear rules and order at work is vital for success.

If you agree −2 points.
If you disagree +2 points.

7 Even after I have decided something, I am always eager to think through a different opinion.

If you agree +5 points.
If you disagree −5 points.

8 I do not like questions that can be answered in multiple ways.

If you agree −6 points.
If you disagree +6 points.

9 I usually make key decisions fast and with a lot of confidence.

If you agree −4 points.
If you disagree +4 points.

10 I can usually see how both sides could be right in most conflict scenarios.

If you agree +5 points.
If you disagree −5 points.

11 It is annoying to listen to indecisive people.

If you agree −3 points.
If you disagree +3 points.

12 I prefer interacting with people who have a very different opinion to me.

If you agree +4 points.
If you disagree –4 points.

13 I often find having multiple options confusing when attempting to solve a problem.

If you agree +1 points.
If you disagree –1 points.

Now mark where you land on the spectrum, from –54 (Hedgehog) to 54 (Fox).

Hedgehogs: You tend to hold on to one or two big ideas to understand the world, how it works and where things are going. You enjoy understanding everything about a subject and simplifying it. You tend to force things or reduce things in order to make them fit into understandable boxes. You express your views with great confidence. You tend to fare better at short-term forecasts but you occasionally get the further-out predictions correct too.

Foxes: You reject the idea that there is one only one model to understand the world, and instead seek out the best approach that fits the problem at hand. You are sceptical of individual theories and enjoy merging them. You tend to adjust rather than force an explanation. You tend to be shy when forecasting and use words like 'however', 'perhaps' and 'more so' when giving your views. You are better at long-term forecasts.

Fear not if you are a hedgehog, it's not a bad thing at all! Tetlock points out that whether you fall closer to being a fox or a hedgehog is largely irrelevant as it is a spectrum and not a final state. The next step is to make better predictions and forecasts now that you know this either way – it's unlikely anyone is 100 per cent either type. Both will get things wrong and both should learn from their mistakes (keep a tally) but also celebrate getting it right. This 'tally' could be a mental tally or a document you keep somewhere so that you can refer back to it – either way works, it just depends how you would like to keep yourself most accountable.

So how can you make better predictions and forecasts?

Forecasting is difficult, time-consuming and requires a lot of resources. All of these elements require one thing – change. This is why leaders of successful firms fail to see changes coming down the track and continue to use the same strategy that has 'worked before' or stick with the dreaded 'because it always works that way'. This sort of thinking doesn't inspire or get businesses to the next level; thinking this way causes wheels to spin and eventually leads to stagnation.

Good questions to ask clients or employees (and yourself) when you first start working with them in this arena include:

- What would it look like if you started over?
- What would you change if you had a magic wand?
- What are the three most important reasons to change?
- On a scale of 1–10 (1 being 'not at all' and 10 being 'business critical'), where does this change land on the 'need' scale?
- What's the risk(s) of doing nothing?

These questions are asked for a variety of reasons but the main one is to understand various perspectives and willingness to change. Once these are being explored in a careful, methodical but fluid way, understanding how to get change happening is significantly easier. Download the question sheet to help yourself ask better questions, give and get better briefs, understand your clients more and get on with friends better.

Great, so what's the problem?

Change is scary and change is the unknown. Change is challenging what is already happening and not being satisfied, which is often a hard place for people to be or to think they can move from. The biggest issue with change is that it often points at things that are wrong, which means blame, and blame is uncomfortable for all parties concerned. In short, change is hard and is fraught with risks.

But it doesn't have to be.

I am a huge fan of exploring change for the sake of change – a lot of people aren't. I enjoy pushing things and buttons (ask my poor family) to see the result and whether things can be improved or what can be learnt. It is important you understand and recognize when people are uncomfortable, as while this can help change along, it can often hinder. Changes in tone, sentence length and body shifts are all signs people would rather be somewhere else or that something is wrong. Be mindful to make people feel safe throughout the change process.

Using equations is a good way to put people at ease – maths tends to put people at ease as they believe in maths and that it is absolute. One such equation I use (and often edit) is David Gleicher's Formula for Change (Beckhard, 1975). Often referred to as 'the change equation', the formula describes what is needed in order for a person, business or thing to change.

The original version can be seen here:

FIGURE 3.1 David Gleicher's Formula for Change

$$\text{Change} = \frac{\text{Dissatisfaction}}{\text{with the present}} \times \frac{\text{Vision of}}{\text{the future}} \times \frac{\text{Clear first}}{\text{practical steps}}$$

In essence, your job when working with clients or employees is to work together to understand what part (or parts) of this equation you or they need help with. Some people are just interested in meeting and networking but most have clear issues in one or two areas when you ask some basic questions.

It would be remiss of me not to mention that the Gleicher formula was modified by Kathie Dannemiller (Cady *et al*, 2014) in the 1980s to include resistance – a key area for companies that is often overlooked and left until the end which usually leads to poor results.

Her formula looks like this:

FIGURE 3.2 Dannemiller's revised Formula for Change

$$\text{Change} = \frac{\text{Dissatisfaction}}{\text{with the present}} \times \frac{\text{Vision of}}{\text{the future}} \times \frac{\text{Clear first}}{\text{practical steps}} > \text{Resistance}$$

D = Dissatisfaction with how things are now

V = Vision of what is possible

F = First concrete steps that can be taken towards the vision

R = Resistance

Resistance is a huge part of change that can kill ideas and movements dead if factors causing the resistance are not evaluated and either removed or mitigated against. Dannemiller has also spoken about the absence of any one element and how this will again impact success – a good thing to remember when applying this formula to your issues or business.

I have found it useful to alter the formula slightly when dealing with specific financially driven or motivated individuals to bring costs to the forefront. This version of the formula – or the HERE/FORTH version if you prefer – looks like this:

FIGURE 3.3 HERE/FORTH's Formula for Change

$$C = D \times V \times F > X$$

C stands for change, D is the level of dissatisfaction with the present, V is the clear vision of the future (or where you want to get to), F is the first practical steps. Looking over my last few years as a young business, 'F' is where most of HERE/FORTH's revenue came from. The ability to identify a clear path to change is often the hardest part for companies to manage without external help (and we will discuss how to overcome this in later chapters). X is the cost of the change. The reason for resistance being removed is simple: resistance can be never ending and usually serves to put off progress. For most clients, the desire to make change and remove resistance has been removed early on in the process. Resistance can also be factored into first practical steps if handled correctly. In essence, this version of the formula assumes a significant desire to change is present or else it is a wasted exercise. However, financial cost is a very real issue – sometimes the desired change simply costs too much and therefore must be taken into account.

Whichever formula you prefer (I find Dannemiller works well with more hierarchical businesses or type-a people), taking into account the way the business works, the way it has accepted previous change and the current mood of employees is critical to any change programme.

Why don't people change?

Sometimes the reasons for resisting change are obvious or overt – the employee fears a power shift, the change will be stressful, new skills will need to be mastered and they will look unskilled to others. Sometimes, though, people just fail to change. I have had this happen with clients' teams and teams I have managed, despite them being completely competent, dedicated and supporting of the changes being asked of them.

There is a lot of psychological research that discusses change management, productivity and decision making, and the conclusion of many studies is simply that deep down (mainly unconsciously) people fail to change because of a hidden rival activity or obligation (such as fear that completing work will lead to harder challenges straight after it). This is then miscategorized as resistance or a lack of dedication to change.

Dedication is a hard thing to create and maintain

This is a key area for this book – your dedication needs to be greater than letting things stay as they are, or worse, letting things slip. The problem with change implementation has big foundations in the way departments are run, hierarchical perception, previous change successes (and often, importantly, failures), recent 'asks' of the employee(s) and the way the benefits are presented. Supporting your employees is easy to say but difficult to do; is making a change that will make your company more money the right message or would 'The results of this change will enable us to take Friday afternoons off during the summer' be better? Direct benefits save money and time for companies but employee benefits aren't as direct or as quick to come so they are 'carrot and sticked' into reality. The smarter way is to understand the motivations of the group, which could be money, but it could be that they want the time to enjoy the money they have already rather than have more money and no time to spend it. Naturally, time off is not everyone's motivator – this has to be judged carefully and re-evaluated on an ongoing basis. Disruptive technologies are a hard sell and they often seem too far out to be of any impact; either that, or other short-term goals might seem to take priority. Change doesn't have to be a big thing – it should be a 'little and often' approach.

So what do I do? I'm not a mindreader...

Change implementation is fraught with issues, be they technological, behavioural or data-related. From experience, change is often too swiftly enforced. Almost overnight, things are altered in a 'was A, now B' way which rarely works unless everyone is on board (which almost never happens, so avoid it if you can unless the change is mission-critical). The key is telling people what is happening, showing them the benefits and getting them involved. Changing things or forcing things rarely works and I have suffered from this approach in the past. The results are not pretty and can be damaging to team morale and productivity in both the short and long term. Instead, managers and senior teams need to really understand the people behind the job title if change is going to last. In order to make it a reality, set yourself goals about finding out what Sandra and Bob are assuming will happen, what else they have on, what conflicts them, how they like to do things and why they do things that way. With the Millennial generation storming into the workforce (which we delve into in Chapter 12), understanding employees at this level will soon be essential.

What happens if change is poorly thought out?

A lot of repercussions can happen when change is poorly thought out and implemented; I term this 'ba-nge' or 'bin-novation', a change that doesn't really need to happen or happens but not for a good enough reason. While previously I mentioned that all change is good, change for no reason can be damaging. Often this sort of innovation comes after a significant period of stagnation or general 'business as usual' and is categorized by a one-off, firework activity like a hackathon, day of learning or similar 'there and then gone' event. These events can work – this is not to say these intensive events can't spark something or help in the short term – but long-term change comes when specific things are in order (such as long-term vision, a clear dissatisfaction with the present and a core understanding of the real issues and limitations involved).

Let's use hackathons as an example. Hackathons are intensely focused vacuum-style events where various people from a number of backgrounds come together to solve a problem. Many brands including Honda, Tesco, General Electric and Salesforce have heavily promoted their hackathons in the past and many will in the future. While hackathons do have many bene-fits (including a hefty amount of quick/easy PR value) they also have several significant drawbacks:

- Innovation doesn't traditionally come from leaps forward, rather it is an iterative process. Certainly, getting out of the day-to-day can uncover huge areas that have been previously unconnected. However, this is rarely seen at hackathons. Instead, because of the time constraints, real-world issues being ignored and adjusted working parameters, simple iteration rather than innovation is often the goal. This is why there is little evidence that hackathons lead to ongoing or long-term market success.

- The vacuum these ideas are created in is often just lip service because of the contextual knowledge and expertise – this isn't the fault of anyone who attends and contributes but it is a core concern if the business running the hackathon doesn't have the drive to push beyond the hack-athon and really make change happen.

The right tools can make people change (or see change) easier

Instead of hackathons, and I'll say again they can be useful, I prefer to use several workshops with clients to elicit new thinking, change and under-standing across the various departments and levels.

SWOB

Similar to the much-used SWOT (Strength, Weaknesses, Opportunities and Threats), the SWOB analysis workshop replaces Threats with Barriers and gets to the heart of individuals' personal styles and areas for development. Beyond simply understanding themselves (and co-workers) better, SWOB analysis enables individuals (with and without help from managers) to create and put in place plans and structures to encourage growth and see change as a methodical process rather than something forced or looming over them.

SETTING

A large room, minimal distractions on tables, plenty of water, Post-its and writing materials. On the walls, a whiteboard or a projector – you may want to put thought-starter posters up or have a concise version of the steps below. The session is about focus and clarity so anything that helps achieve that is welcome.

Time required: 1–2 hours.

Max group size: 30.

FACILITATION

The leader must be above average in presenting and speaking in front of a group. An external individual is highly recommended as this task can be uncomfortable for participants.

MATERIALS REQUIRED

A4 paper or Post-it notes.

Pens.

Space for contemplation/working and group-work (key to separate these areas).

RUNDOWN

1 Individually, all participants take a piece of paper and list out their strengths, weaknesses, opportunities and any barriers that could stop them changing, improving their weaknesses and acting on opportunities. (Duration: 20 mins)

TOP TIP: Get participants to write down as much as possible – use cues to spur on deep thought. For example, 'Think of a time when you got something right/wrong – what made it a success/failure?' 'What was written in your last end-of-year review?' 'What would someone reviewing your last project say about how you did it?'

2 Using the SWOB analysis just completed (some areas will be fuller than others), now get participants to think about three to five areas they want to develop. While you can do this in small teams to spur on ideas, it is mostly best if done by the individual alone.

TOP TIP: Again, cues can be good. Think about the issue at hand – a change or new implementation. Ask participants to think about what they want to get out of this change. Ask for and give permission to give brutal honesty.

3 Individually now, participants get to rate their areas for development using some or all of the following questions (40–60 minutes):

- Using a scale of 1–10 (1 = poor, 10 = could not be better), where would you rate yourself in this area?

- Using a scale of 1–10 (1 = same place, 10 = finished), give each of your areas for development a score for the next 30 days, 90 days and 180 days.

- What could you do – or what do you need – to accelerate your own development?

- How will I overcome the barriers I have noted in order to make sure I hit my goals?

- What are my first practical steps tomorrow, next week and next month?

- How will I know that I have succeeded or am on my way to succeeding? What will I see? What will I be able to show or demonstrate?

4 In groups of three or five, participants then take turns in sharing their development plans with the other members of the small group. Participants should be pre-prepped to ask each other questions that help people push and question their thoughts. Remind all participants this is a safe space and everything is aimed at helping one another. Open questions that focus on refining and solidifying ideas or issues are best in this section:

- What were the most important factors in that point?

- How exactly do you think you should start?

- What led you to that conclusion?

- Who benefits most from _____?
- What types of things have worked/failed in the past that are similar to what you are suggesting?
- What was your previous experience(s) like doing _____?
- What would a bigger goal look like?
- What is the likely outcome if you don't do _____?

5 As a group, participants can also ask if there are other areas to be developed – it's ok of course if this answer is no but make sure everyone knows that leaving things out rarely helps anyone move forward.

6 Individuals now finalize their development plans, asking for any other assistance needed or making notes for things to do immediately, next week etc.

7 Ask the participants how they felt about the task, what they learnt, what they think about their goals and any issues they foresee in the next 48–72 hours.

Forecast tape

The Forecast Tape exercise is a visual way of getting employees (or whom-ever) to think about different parts of themselves, the consumer, the business, life and the future. It is based on the Diffusion of Innovations curve proposed by Professor Everett Rogers (2003), which explains how and why different ideas and technologies spread through cultures. Diffusion is the process by which an innovation is communicated through certain channels over time among the participants in a social system. Rogers proposes that four main elements influence the spread of a new idea: the innovation itself, communi-cation channels, time, and a social system. The five categories of adopters are innovators, early adopters, early majority, late majority and laggards. Diffusion manifests itself in different ways in various cultures and fields and is highly subject to the type of adopters and the innovation-decision process. The exercise is about high energy, deep thought and openness but the fun nature of the exercise means these elements come naturally. It's worth doing this exercise twice a year to see where people have moved to.

FIGURE 3.4 Diffusion of Innovations

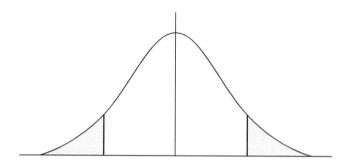

SETTING

Get the team together – have them stand in front of a blank wall and explain the theory to them. Now, as a group, the team has to draw the curve onto the wall using the thick masking tape – everyone gets a piece to put up (alternatively you can set up the tape before the group gets there if time is short). Once the curve is complete, step back, admire the curve and proceed with the task rundown below.

Time required: 1 hour (or less if no setup).

Max group size: 30.

FACILITATION

Lead needs to understand the theory fully, be lively and get everyone involved.

MATERIALS REQUIRED

Thick masking tape – alternatively a projector can be used.

Large Post-it notes.

Large wall or window.

Pens.

A4 paper.

RUNDOWN

1 It is important to explain to all participants that the exercise is more about self-reflection than comparison or team dynamics. Remind people about the theory, review the categories of adoption below and ask participants to

think about where they see themselves and more importantly the customers of the business (be as specific as possible):

- **Innovators.** These are confident risk takers who often have the highest social status, are financially more secure than other groups, are social and cultivate relationships with science communities and seek interaction with other innovators. Due to their tolerance of risk they adopt many technologies that may ultimately fail but their financial security (and networks) mean these failures impact them less.

- **Early adopters.** These individuals have the highest degree of opinion leadership among the adopter categories. More social than late adopters, early adopters also have higher social status and financial security but are more discreet than innovators when it comes to adopting technology. Early adopters judiciously choose which technologies to use to help them maintain or increase (or further solidify) their position as a hub.

- **Early majority.** The early majority follow the innovators and early adopters but only once it is safe or obvious to do so. Early majority have above-average social status, connect with early adopters but rarely hold opinion leadership positions within friend or work groups.

- **Late majority.** The late majority adopt innovations even later due to a high degree of scepticism even after the majority of society is using the innovation. They are often categorized as having average social status and money is a factor in most decisions. Their social circles are likely to contain others very similar to themselves and are unlikely to contain opinion leaders.

- **Laggards.** Laggard are the last to catch on to or use an innovation. This group has very little to no opinion leadership but often features a high aversion to people looking to or speaking about change. Laggards typically tend to be focused on 'traditions' and 'traditional values' and keep to friends and family for advice.

- **Leapfroggers.** When resistors upgrade, they often skip several generations in order to reach the most recent technologies.

2 After answering any questions that participants may have about the categories, ask participants to ponder where they lie on the curve if you ask the following questions.

At work when was I:

- An innovator?

- An early adopter?

- The early majority?

- In the late majority?

- A laggard?

You may want to provide an A4 version of the curve for people to write personal notes or thoughts on. The sheet of paper is also a great way of creating a reminding device for the participant to keep and refer back to. Spend about 10 minutes on this.

3   After around five minutes, pair up individuals and get each pair to create their own curve using the large Post-it notes or A4 paper. Each participant must share what they wrote down for each section but they do not necessarily have to go through it in any order. Once they have their curve ready, have each participant reflect on how they differ from (and are similar to) each other. Bonus idea: photograph each participant standing beside their curve and send the photo to them three days and six months after the task to act as a reminder and thought-starter.

From experience, the best questions to spur on deep thought about the differences are:

- What area surprised you the most?

- What area surprised you the least?

- What differences did you and _____ have?

- Knowing what you know now, how will this information change the way you think about _____ and _____?

- How do you think being a _____ impacts the way you approach your work?

- Is there anything you think would benefit from being different? How could you make it different?

Bonus idea: You might want to take a pre-poll from senior execs about where they believe your current, ideal and new customers might be on the curve and then compare it to where participants think each segment is. Then, as a group, participants move to where they think the execs said and – after the true answers are revealed – then reflect on and discuss the results as a team. Now, go a step further and really brainstorm what technologies and innovative products they may (or may not) use to best serve them moving forward.

4 Finally, after everyone has gone through their own curves, all participants gather in front of the curve and some insights can be shared – by the facilitator and participants. (Note: Participants sometimes find it easier to be called on in group situations so make sure you note down who had interesting ideas.)

New product brainstorm

A mash-up is a collaborative idea-generation method in which participants come up with innovative concepts by combining different elements together. In a first step, participants brainstorm around different areas, such as technologies, human needs, and existing services. In a second step, they rapidly combine elements from those areas to create new, fun and innovative concepts. Mash-ups demonstrate how fast and easy it can be to come up with innovative ideas.

SETTING
A large room with clean walls. The session is about connecting dots, seeing connections and thinking differently – the more creative and busy the space is the better; it should be a high-energy exercise.
Time required: 1–2 hours.
Max group size: 20.

FACILITATION
Due to the activity level required, the lead should be ready to spur people on, dip in and out of various areas and able to jumpstart groups. An external individual is highly recommended as this task benefits best when all levels are involved.

MATERIALS REQUIRED
Wall space – lots of it!
A4 paper or Post-it notes – lots of them, multiple colours, sizes, shapes – the more the merrier.
Pens.
Stickers (three to five times the number of participants).

RUNDOWN

1   Set up the session, its aims and objectives and then get down to business. Often referred to as the Post-it note ravaging stage, the first part of the task is to complete a group exercise of shotgun brainstorming. Simply ask the group to brainstorm around three areas; it is often best to do these systematically but it is not necessary. It is important to keep all the ideas for each area separate for now but also that the session is fun, lively and as fast paced as possible (playing fast-paced music helps). If it is a big group, you may want some people to help stick things up on the wall but make sure everything gets captured and no idea is forgotten. The three areas to brainstorm around are:

   - **Technology.** Have the group shotgun down all the types of technology they can think of – get them to think about what they use during the day, when they are away from the office and other scenarios.

   - **Needs.** Specifically needs that people have like good sleep, security, connection – some may seem basic but all are valid. TOP TIP: It may help to keep a copy of Maslow's *Hierarchy of Needs* around.

   - **Services.** These can be apps (e.g. MyFitnessPal), platforms (e.g. Meta) or games, photo manipulation tools – everything that already exists.

2   Once you have the board covered, remove duplicates and clarify anything that requires it. After this is complete (don't take too long, keep the momentum up), either randomly or in predetermined groups, organize participants into groups of five or less. Instruct them that the next 15 minutes are to be used to come up with as many concepts as they can. A concept is an idea that combines either two elements or one of each of the areas just brainstormed. Do not take anything off the board but have the groups stand up and create new ideas by combining elements. Inform the teams that each idea must have a name and to write the idea simply (i.e. either two or three areas) on a single sheet of A4 paper, one idea to one sheet. Make sure the timekeeping keeps the pace up.

3   After the 15 minutes of fervent activity have finished, it is time for the groups to present the new concepts. Each of the concepts is put up on a fresh wall so people can see the work that has been created by the entire group. Keep this process speedy and make sure people remain upbeat and non-judgemental. All ideas are good ideas in this process, whether serious or funny – everything goes up and everything gets heard.

4 Lastly, hand out either three or five stars to each participant to choose their favourite idea or ideas. Participants can choose to either support one idea with five stars or five ideas with one star or another combination – participants are allowed to vote for their own idea.

5 Now you have two options: you can end the session there and walk away having created a new idea that can be evaluated at a later stage or by a different group (a board etc), or you can hunker down and – still as a group – spend 20–30 minutes working up the idea into a viable proposition. This involves discussing the idea, the thought behind it, resources needed, business models and how it would actually work. Sometimes this process can get heated as conflict can occur but capture it all and feed back to the group what will be done next (will it be taken to the board, evaluated by outside counsel etc?).

6 Whether you decide to end the session before this or after it, make sure you fully debrief the participants, reminding them why you did the session, what you did and what happens next. You may also want to ask participants a few questions or have them fill in a short feedback form:

- What was easy? What was hard?
- What did you learn about yourself?
- What did you learn about your team?
- Pick a person and tell us something they did well during the session that you'd like them to know.

Note: This workshop exercise can also be used with open data sources, partners or big issues – each of these areas can add new dimensions to the exercise but also may enable further discussion or refinement of the idea once created.

Conclusion: innovation needs a flexible framework

Hopefully, both the preamble and the workshops have highlighted the need for a flexible approach to forecasting, prediction and making decisions about changes needed. Additionally, both highlight core pillars that require consideration in order for good decisions to be made: technology, behaviour and data. This is important if needs (and not wants) are to be met stemming from the products and businesses of tomorrow; the businesses that not just survive but thrive or spring up because of the changes going on in the world.

It is for this reason the TBD framework came into being and exists in its current format. The following chapters look at the TBD framework in more detail and will enable you to:

- increase your preparedness to handle the unexpected;
- increase your evaluation skills when it comes to disruptive technologies;
- remain calm when changes happen, because you have a system to rely on;
- make clear and well-thought-out decisions based on a replicable tool.

04

The TBD framework: an introduction

In this chapter, we will look at the simpler version of the TBD framework, why it is useful, what it offers businesses, how it originated and how to apply it to your business or department. It is important you understand the why behind most things but more so with any framework you choose to adopt, as you want to know the thoughts behind each element and the process by which it came to be to make sure it is a fit for you. Subsequent chapters go into greater detail about the advanced version of TBD and what the framework might look like in the future.

The future needs to be agile

The need to be agile when we live through increasingly uneasy and changing times isn't a hard sell to most businesses. A quick glance at any news outlet tells us there are issues ahead for most businesses beyond simple competition (regulation, legal changes, political instability). Businesses (and individuals) need to be flexible in order to remain less pregnable to being knocked over. Agility means being able to move quickly and easily but I also add being resilient to attack – one or two hits should not push your business over. Beyond simply protection and risk reduction, being agile enables a different mindset to spread throughout organizations and allows different ways of thinking and producing new and exciting ideas.

Financial markets are in flux and a general sense of unease has taken over society thanks to terrorism, leaks, privacy issues, the internet and lack of faith in most governments. Technology does a good job too, be it cars that drive themselves, robots that 'want' to take your job, the medicine inside you (that you swallow) or computer viruses that aim to start nuclear warfare. It is easy for people to think things are moving in the wrong direction, partly due to communication of what's happening but also because so many times things are retracted, changed or flip-flopped on.

It is for this reason that disruptive technology is so interesting and can be so spectacular when it comes almost out of nowhere. When a disruptive technology happens, it sends shockwaves and often you can't predict what else will be disrupted beyond the obvious. To some this is exhilarating while others find it disconcerting.

The right mindset is key

As we have seen in previous chapters, disruptive technologies are by their nature highly unpredictable. Companies and technologies that disrupt are often not seen until they are right in 'the rear window' despite us understanding the need for and the value of looking out for them. In essence, a leader needs to ignore naysayers and forge ahead towards the brighter future.

The difference between companies that do great things and those that simply follow the roadmap laid out for them is mindset. John Gardner, a famous leadership expert, calls this 'tough-minded optimism' (Gardner, 1990):

> I can tell you that for [organizational] renewal, a tough-minded optimism is best. The future is not shaped by people who don't really believe in the future. Men and women of vitality have always been prepared to bet their futures, even their lives, on ventures of unknown outcome. If they had all looked before they leaped, we would still be crouched in caves sketching animal pictures on the wall.

I use Gardner's quote frequently when working with large and smaller clients alike. Whether working with large or small companies, the most successful have been the ones that understand the ideas and sentiment around this quote and have chosen to forge ahead even with red tape and corporate barriers in their way. Creating a robust mindset is a combination of original thinking and resilience in the face of big changes but deep conviction is the defining attribute you need to succeed. TBD aims to help a little with all three.

How to toughen your optimism

There are four questions that you need to have clear answers to in order to develop your optimism. Asking yourself these questions will enable you to become more positive about the future and focus your efforts so you are 'tough-minded' when things don't go smoothly or challenges crop up.

1 **Do you have a mission for your business that enables you to stand for something unique and enables other people to believe in what you stand for?**

What are the ideas, ideals and beliefs your company wants to be known for? How do you demonstrate these through your products, marketing, employee relations and corporate activities? Simon Sinek is famous in business circles for his book *Start with Why* (Sinek, 2011). In the book (and famous TED talk: http://bit.ly/DTsinek), Sinek discusses the differences between Microsoft and Apple when it comes to 'the why' behind the brands and how they sell – one (Microsoft) sells with a clear functionality perspective whereas the other (Apple) sells in a more philosophical way, offering a 'way of being' versus functional attributes. In essence, you must agree with their philosophy to buy a product. It's a robust theory and one that many brands do not pay enough attention to at all stages of their business planning.

2 **Do you care more about things than others? How different is this compared to your colleagues?**

Question one is about how you think about things whereas this question is about feelings towards things like co-workers, customers, suppliers and how your organization does business. A good exercise is to create a scoresheet with questions at the top that relate to these things and then rank yourself out of 10; repeat the task every three to six months and you'll see where you are weaker.

3 **Are you as consistent as you are creative?**

This question focuses you on the decisions you have made before. Sometimes people change too much and flit from thing to thing. While such behaviour can be a positive thing that some thrive on, multiple changes (especially in a short time period) can cause uncertainty within employees. During good times and bad times, having consistent priorities and relaying them back to all employees is key to not only being a tough-minded optimist but spreading it as well.

4 **How does your company's history help its future?**

Great leaders don't ignore the past. Instead, these individuals understand how a company got to where it is and reconstruct what has happened into new actions. This does not mean ignoring new breakthroughs for the sake of retaining old ways but it means evaluating each and forging ahead with the past in mind.

Thinking about these questions (there are never firm answers) will enable you to strengthen your own tough-minded optimism about the future. Without this optimism, you, your colleagues and your organization won't be able to fully conceive the bright, shiny future that is possible.

The forecast is still cloudy, fast and changeable

Any trend or forecasting agency will tell you that looking into the future is difficult. Companies such as these spend time looking at indicators, data sets and speaking with experts about possible outcomes but as we have seen in Chapter 1, they are not super beings or people with special powers or particularly elevated IQs – they have a formula or a way of working that works for them and it will change depending on the topic they are looking at.

This is exactly how TBD works; a flexible framework that enables you to formulate a practical plan with regards to disruptive and emerging technologies so you can move forward with the right amount of attention to each.

The origins of TBD

TBD was formed in my early twenties at a 'party' in Los Angeles. It was the usual sort of thing – a party that isn't really a party because it's at the end of the work day and no one is very comfortable at this sort of event (imagine a meeting with alcohol and lots of nodding and you get the picture). Shades firmly on thanks to the blazing sunshine, I walked to the edge of the house that overlooked Los Angeles and began thinking about how it all works as cars, planes and the world continued on about their business. 'It's a great view, isn't it?', a voice boomed to my left. 'It certainly is', I responded, 'It's easier to see how it all fits together from up here.'

The man – in his late thirties and well put together – turned out to be a young agent for William Morris who was 'expanding his network'. We continued chatting for about five minutes with the conversation even getting quite heated in places about the way Hollywood was changing and what was driving these changes. Technology was at the heart of it. I remember saying to him, 'The trouble is that you have a willing audience who wants things faster, cheaper and more pleasurable. The cinema is increasingly failing consumers on all of these. It is only logical that there will come a point where movies will go "day and date" because theatre numbers and people will demand it.' The agent scoffed of course; he believed that the power lies with the studios (and he is of course correct to a degree) but more so he felt that the numbers didn't add up – i.e. a movie could never make back its money unless it had a three-pronged 'attack' as he put it – box-office revenue, home entertainment sales and long-tail distribution.

Nowadays we know better. Movie studios reduce risk and make money in multiple ways so return on investment is easier to generate despite falling cinema attendance in multiple major territories. While the formula continues to change, no one can deny the likes of Netflix, iTunes, Apple TV and Amazon are all disrupting the film and content business. Sean Parker, co-founder of Napster, the music sharing service, believes the days when movies can be streamed on the same day and date they open in theatres are closer than the studios want to admit. The COVID-19 pandemic certainly has closed the theatrical gap but whether it stays that way is still up for debate. People abhor paying the prices and also the 'stress' the theatre can be.

Back at the party, and with my colleague still nowhere to be seen, the sun was setting and exits and eyes were beginning to be made by the guests. The agent and I were still semi-arguing about the rise and power of technology in transforming his 'sacred' business, from creating stars to illegal distribution of content. Time and again, technology and people came up but this time the discussion turned to numbers – the agent had not realized the numbers associated with things like piracy, data storage and data transfer rates. In essence, his reliance on old ways and the pillars of power had blinded him to potential new ways and more importantly new revenue streams. The evening was drawing to a close and quite flippantly I remember saying, 'Look, everything really boils down to three things: can they do what you are asking, will they do what you are asking and will enough of them do what you are asking?'

I wrote down the words can, will and enough on a napkin that had other scribbles on it but I remember drawing lines to the words technology, behaviour and data and TBD was born. Actually, let's say it was fertilized, as TBD has been adapted over the years into two versions: the simple and the advanced.

The short version is the napkin version adapted into a process that is described in more detail below. In essence, this is a shorthand to understanding the need for more research and insight gathering on a 'change'. The advanced version is more robust and takes longer to complete but provides the users with a plan for the business, not just an answer to (or course of action for) an immediate issue. Both are designed to be used separately or in conjunction with one another.

Why two versions?

From working for myself in-house at Myspace for large agencies and boutique firms, I know one thing unites them all: time is tight. Even with the best will in the world, time is a finite resource and there are multiple calls on it. Often 'good enough' has to do, but at regular intervals you have to put a stake in the ground and focus: TBD enables you to do this. The short-form version is an everyday shorthand that enables you to look at issues, platform feature changes and new platforms quickly whereas the longer version enables a calculated analysis of the landscape the company or brand finds itself in. Both have drawbacks but equally both have significant benefits when used properly.

The short (or simple) version of TBD is essentially reductionist in nature. In other words, the result is through careful removal of information in order to determine a good course of action. John Maeda, ex-MIT head and now analyst for KPCB, has always been fascinating to me; he has dedicated his life to design and how it affects people and now specifically the business world. I have seen Maeda speak several times and his way of working means you always learn something new. In his book *The Laws of Simplicity* (2006) Maeda makes the argument for 10 independent laws that help to make anything simpler. I used these when refining the original TBD into what it is today:

1 Reduce – the simplest way to achieve simplicity is through thoughtful reduction.

2 Organize – organization makes a system of many appear fewer.

3 Time – savings in time feel like simplicity.

4 Learn – knowledge makes everything simpler.

5 Differences – simplicity and complexity need each other.

6 Context – what lies in the periphery of simplicity is definitely not peripheral.

7 Emotion – more emotions are better than fewer.

8 Trust – in simplicity we trust.

9 Failure – some things can never be made simple.

10 The one – simplicity is about subtracting the obvious and adding the meaningful.

These 'laws' can be used beyond the TBD process in the strategy creation process as a whole. 'The devil is in the detail' is not a quote many can subscribe to when it comes to disruptive technologies. Instead, people usually prefer a 'don't ruthlessly plan; ruthlessly execute' approach and these are often the words you hear when executives are asked why they are successful in profiles in outlets like *Fast Company*, *Entrepreneur*, *Forbes*, *Fortune* and *Harvard Business Review*. The old George S Patton adage seems to be true: 'A good plan violently executed now is better than a perfect plan executed next week.'

Initially, TBD was simply a list of questions that a person went through in order to explore an idea more fully. However, it developed as I used it and more people saw it in action.

What 'simple' TBD is set up to do

The Introduction and Chapter 1 taught us that things are speeding up and that the types (and pace) of technological change are moving faster than ever before. Often people feel out of their depth and suffer from analysis paralysis, death by PowerPoint and binding by red tape. 'Simple' TBD was set up to acknowledge and combat this by answering one simple question: 'Should we do more research into _____?'

Before you jump into applying the simple version of TBD it is important to point out a few caveats:

1 Asking and answering questions is important but can be misleading. TBD relies on asking and answering a variety of questions in order to move forward with a clear issue but asking 'Should I do A or B?' could be a false dichotomy. This question implies that A and B are mutually exclusive (i.e. you couldn't possibly have both options). While simplicity is the focus and key of TBD, the options it gives you are essentially binary and as we all know, business is not always so simple. Using the 'simple' TBD you may miss options, which is why the full TBD framework was developed. In other words, just asking/answering questions is essentially the wrong thing to do and may not be entirely helpful in every scenario and industry; however, in many instances it can help to clarify some good next steps.

2 'Simple' TBD does not take into account how your company applies or thinks about change. This is massively important if you are looking to

push through new ideas. Just because you think you should do something does not mean you can or will. Understanding this is key to making TBD really work at a corporate level, not just for low-level decision making.

3 'Simple' TBD assumes resources are infinite. 'Simple' TBD is aimed at fast answers to (often but not always) complex problems. Again, business is not always like this but the successful 'fail fast' mentality of Meta, Apple and start-up companies is testament to the world we are living in and moving into. TBD assumes that if five things come along and all are deemed worth doing then a company should, could and will do them. This is unrealistic but TBD remains a way of minimizing these risks early on and helps companies and individuals focus on the current landscape. This issue is addressed in the advanced version using a ranking and prioritization methodology.

Even with these caveats, simple TBD can be incredibly useful for prioritization, further research decisions, resource allocation and quick wins. All businesses are looking for frameworks, formulas and methodologies they can use to streamline and optimize their strategies in order to save money or time. As with most things in business and life, it may not necessarily be the thing you do that gets you the most value; it may be the value that thing brings you.

The 'simple' TBD framework

Step 1: Create your Decision Matrix

Based on a Risk Matrix design, the first part of the 'simple' TBD framework is to create a matrix of outcomes or a 'Decision Matrix'. In other words, this is a list of what you will do if and when an idea, technology, new product, feature or platform emerges or changes in some way. These outcomes can and will be different for companies and industries in the same sectors but the point is to make it unique to the company at hand. The Decision Matrix is a way of pre-assigning outcomes to specific numbers so that when you finalize the TBD score you can go back to the matrix and see what the next steps are. The key is to make them as specific and action-oriented as possible. No wishy-washy answers are allowed in any Decision Matrix.

The 0–10 scale was used as 10 is an easy number to understand and is commonly used to judge things – some clients prefer to use 0–100 but 0–10 usually simplifies things. Zero may not necessarily mean 'no action' but it is usually where most people assign the least activity, importance or instructions. Start with a blank piece of paper and list out what would be the decision if something was ranked 30 on the scale, based on a 0–10 rating for each of the three TBD questions outlined in step 3 – in other words it could not be a more perfect fit for the company based on its current aims, goals and objectives. Once you have that, write it at the top of the piece of paper.

Some possible examples of best-practice decisions that have come from clients include the following:

- Within 24 hours, write a 200-word recommendation plan for the senior team to vote on (majority approval means the budget of $_____ will be applied and a specific test plan will be laid out and executed within 48 hours).

- Within 24 hours, the senior team will meet and discuss the issue at hand. At the end of the meeting the group will vote on the next course of action from three options: go ahead and fund the test case, approve further research, or deny further action at this time but revisit in 3/6/12 months.

- If the cost of a test would be under $_____ then [MANAGER] is approved to move ahead with testing the issue. If more than this sum is required, [MANAGER] will seek approval from the senior team via a vote within 24 hours after an email case has been submitted.

What would a zero rating cause you to do? This may not be anything or it may be a 'wait and review in six months' decision. Whatever you choose, you now have the best- and worst-case scenarios. You have desired outcomes for both the best and worst of cases. The next step is to start filling in the points in the middle of the Decision Matrix. This step is key, as experience and the law of averages tell us that a lot of technologies will fall in the middle of the Matrix. Spend some time really examining what you will do in the case of a 15 or a 17 – where will you draw the line from one decision to another? Think about the benefits to the company and what the risk of not doing something might be – this (and other questions) can help move things up and down.

Now think about when you would want to make a separate distinction to the best-case scenario (i.e. 30) you just wrote down. For some people this will be 29 and for others it will be closer to 25. The number itself doesn't necessarily matter but allowing yourself time and space to understand the difference between a 25 and a 30 is key to making decisions that could cost or save companies thousands of dollars. This process creates 'bands' or areas where specific decisions will be made enabling you to make fast determinations on the course of action that a technology, feature or upgrade proposes.

The final step is to fill in other points in the Decision Matrix. What happens at 5, 10, 15 and so forth? Some companies will have one decision for every number (although this isn't recommended) while others stick at three (a traffic light system) or four. The most successful companies that use TBD create a few bands for decisions (i.e. 10–15) with specific simple actions for the lower bands and intricate instructions for upper bands rather than lots of complex decisions for each number or lots of numbers. Having fewer

FIGURE 4.1 Risk Matrix

Likelihood	Likely	Medium risk	High risk	Extreme risk
	Unlikely	Low risk	Medium risk	High risk
	Highly likely	Insignificant risk	Low risk	Medium risk
		Slightly harmful	Harmful	Extremely harmful

Consequences

FIGURE 4.2 Decision Matrix

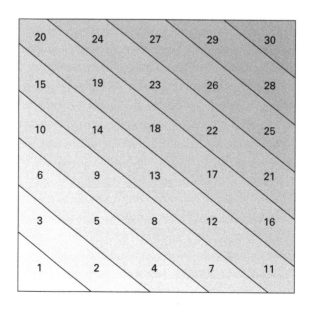

but more specific bands enables you to move faster and be more efficient and often causes fewer misunderstandings, as the same result is often achieved even if the T, B or D scores become disputed.

The Decision Matrix is based on a typical Risk Matrix, which is used to determine the response when something bad happens (or to help avoid riskier strategies).

The Decision Matrix is slightly different because we do not assign likelihood or severity of disruption to it – it is simply a score to determine a course of action.

TOP TIP

The best decision matrices are ones that are not created in a vacuum. Grab a team of people together – all levels is best – and talk openly about best cases, likely cases and how the company has dealt with similar things in the past. Be open and honest about budgets and time requirements and agree the wording, the deadlines and the action to be taken. Doing this will ensure several things, from compliance and understanding to encouraging future collaboration.

Step 2: Decide on the questions you want answered

This can be challenging as often businesses aren't sure of disruptive technologies and rapid changes as their structures are inflexible or change requires a lot of energy. However, done correctly and with practice, forming these questions can often be done quickly and efficiently, either ruling out further investigation or beginning a good course of action. Naturally, the questions differ and the phrasing changes from company to company but essentially you are looking for a 'What should we do because of _____?'

Most of the time the questions will look something like these:

- What should we do about _____?
- Should we start using _____?
- Will _____'s new feature, _____, help us make more money?
- Will _____'s new feature, _____, help us save time?

It is important to remember that simple TBD is not a cost–benefit analysis or anything similar, but a quick way to decide what to do next. In essence, TBD is a motion giver; you either need to move forward or you don't (and perhaps decide to revisit or check back in the future).

Step 3: Use TBD to obtain the TBD score

You are now required to make some judgements on the elements of TBD (Technology, Behaviour and Data) in order to determine a score that relates to your Decision Matrix.

1 TECHNOLOGY (THE 'CAN THEY' ELEMENT)

The question to ask yourself is: based on the evidence I have in front of me, can the end users do what I need them to?

This is often the most important question as it sets the framework for what you are asking about in general. People can be made to do things (behaviour) and in enough quantities (data) but primarily you need to know what it is the thing does and how it works; in short, as much about it as possible. Clients that use this method tell me they usually spend about 20–30 minutes on this section but can spend up to an hour or more if it feels like a big issue or information is scarce.

Once you have completed the research, give yourself some time to reflect on what you have found – read through it, highlight positive and negative points but form an opinion based on what you have found; you may decide that more time is needed so continue your research until you are satisfied. Sometimes with disruptive technologies very little will be available but being aware of this upfront can often help create specific scenarios at step 1.

A score of zero means the users will not be able to do what you are asking – for example, if you want people to be able to post video and the site only allows you to write blog posts for revenue share, this would result in a zero. However, if the blog site does allow people to post video, a score of five or above seems fair because although you can post video, it is not the primary function of the site. Be ruthless with your scoring – refer back to it frequently, challenge yourself on your assumptions and look at the research you have conducted.

FIGURE 4.3 Can they?

TOP TIP

Create an Evernote account early on. Evernote is an online bookmarking tool that enables you to save whole webpages and search within them later on. It also enables you to forward emails to it – it is like your own personal, private knowledge repository that you can access from anywhere with an internet connection (and offline if you choose that option too). By tagging good information every time you find it you will be able to easily relocate it, update it and keep topic notes for future use. For example, you may want one note for 3D printing and another note for 3D-printed fashion or you may combine the two. Notes can then be shared and co-created if there are several people researching a topic at once (recommended).

Sites that help with Technology research:

Academic search engines   Wikipedia keeps an up-to-date list of top journals that can help find unique details and aspects a lot of other publications miss. (https://en.wikipedia.org/wiki/List_of_academic_databases_and_search_engines)

Google Advanced Search   This part of Google is often ignored by businesses because the first page of results usually locates something of interest that pushes them down an exploratory search path. Instead of simply using google.com, use the following link to find a page that will further refine your search to include and exclude certain things. (https://www.google.co.uk/advanced_search)

Industry-specific search engines   Industry-specific or 'vertical' search can be done in a number of ways but the easiest (and often most up-to-date) way is using a specific search engine or database – the Search Engine Guide is an excellent resource to start your research for highly specialized information that is often not openly available. (http://www.searchengineguide.com/searchengines.html)

Noticeboards/board communities   These tools can be rich sources of experts – often people at top technological firms contributing anonymously or obsessive types who enjoy finding and exploring topic areas. Examples of

FIGURE 4.4 Will they?

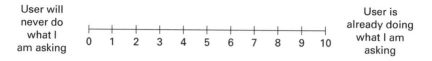

such sites include Reddit, Quora, Voat, Snapzu, Stack$ity, Digg, Hacker News, Product Hunt and Slashdot. A top tip is also to look within these sites for unique and dedicated sites for things like specific Microsoft products or 3D printing communities, or on their own using any major search engine.

2 BEHAVIOUR (THE 'WILL THEY' ELEMENT)

The question to ask yourself is: based on the evidence I have in front of me, will the end users do what I am asking?

Using your research, or conducting more if (or as) required, you will be giving a score to the Behaviour part of TBD. In this section, you must know your consumer (or the consumer that is using the technology) and decide whether it immediately overlaps, or will or could overlap in the future.

Sites that help with Behaviour research:

GWI   GlobalWebIndex is a survey database that tracks how people use the web with an impressive level of detail. Updated quarterly, the service covers 34 countries (546 local regions) and offers brands and individuals granular information on behaviours, device usage, attitudes, lifestyle, demographics, social media usage, app usage, commerce preferences and marketing touchpoints. The database is not read-only, however; the beauty lies in being able to slice the data lots of ways. Some data can be used for free but the paid-for service is well worth the money. (https://www.gwi.com)

Cognitive Lode   Created by Ribot, Cognitive Lode takes the latest behavioural economics and consumer psychology information and distils it down into nuggets of information that can be easily applied. Sign up for the newsletter to get new information delivered to you each week. (http://coglode.com)

Melissa Data   This offers a wide variety of data pertaining to geographic, socioeconomic and identity information that can be accessed for a price (some is free but limited to a specific amount every day). Useful and timely information can be gleaned from the site as it is regularly updated. (http://www.melissadata.com/lookups)

Government sources Both the US and UK governments lead in open data policies with both providing multiple datasets and tools to find out about what the public uses, does and believes on a wide range of topics. Sites can be difficult to navigate due to the large volume of data and different types of information contained within the sites but stick with it and go beyond obvious searches to find nuggets of data and insight.

TOP TIP

Don't forget to look at census data as a jumping-off point; health data is often given its own deeper site by most governments so search for that type of information. (US: www.data.gov and UK: www.ukdataservice.ac.uk/get-data/themes/health)

Facebook Graph Most Facebook data is private but a vast amount is not – when you think Facebook has more than one billion users logging in every month you can get some robust data and insights. Facebook Graph is the tool you can use to query large amounts of information that users are happy to share (or don't realize they are sharing).

TOP TIP

Look at Facebook IQ for more tips and deep data analysis that has been compiled and released by Meta. (facebook.com/business/foresight and developers.facebook.com/docs/graph-api)

---

3 DATA (THE 'WILL ENOUGH' ELEMENT)

The question to ask yourself is: based on the evidence I have in front of me, will *enough* of the end users do what I am asking?

The Data section of TBD is the most useful in determining a course of action as it can often polarize a decision because of the evidence available. In this section, you need to assign a score to this question based on the research you have conducted (or will conduct if more is required).

Sites that help with Data research:

Website and application measurement services   Sites and services like comScore, Nielsen and Compete all measure traffic, demographics and other metrics that show who is using websites and apps. A quick Google search will find any of these sites and the free content they provide. Several give out a wide variety of data and reports on topics via their blogs although almost all have paid-for subscription models that offer readers a lot more data.

Data.ai   Mobile and smartphone data is increasingly important as more and more time is spent on these devices and in places where old technology used to rule. Data.ai is the leader in this field with few to rival the data it has on the various app stores and how people are using their phones. Beyond Data.ai's useful (and regular) free content, they also have a paid solution that enables users to gain granular information and data on market insights, forecasts and changing consumer behaviours. (https://www.data.ai)

Analyst firms   The top large firms are McKinsey, Forrester, Gartner, IDC and IHS but a quick search will find more specific ones too. Understandably, these knowledge-worker establishments don't give out much for free that is totally up to date but what they do give out is often forecast in nature or

FIGURE 4.5 Are there enough?

helps show trend information. Often the work is of a high quality and sufficient for this level of reasoning although more detail will likely be needed should further research be deemed necessary by your Decision Matrix. In this case, you may need to purchase full reports. You should also look for smaller analyst firms that specialize in the areas you are looking for as these will also produce interesting insights and are often more forthcoming with free insights than larger firms.

Economic data   FreeLunch is a great site that offers up high-level economic data (mostly for free). Run by Moody's Analytics, FreeLunch provides historical and forecast data at top and low levels and represents over 93 per cent of global GDP, covering more than 180 countries. The FreeLunch database contains more than 200 million financial, demographic and consumer credit data points and adds roughly 10 million each year. (https://www.economy.com/indicators)

Media and press sites   While obviously likely to contain some sort of bias (as they are often written by PR and marketing departments), press and media sites on platforms and company sites offer a wealth of information that is usually dumped on specific areas that can be searched and accessed through indexes. These vary in position but are often right at the bottom of sites in the navigation bar.

Data USA   Made in partnership with some serious brains who work for MIT, Deloitte and Datawheel, Data USA is a relatively new site that was created to help people navigate the massive amount of public US data. The data is visualized using a number of methods and is free to use. Best of all, the software code is open source, meaning you can take it and develop your own products using Data USA data mixed with your own. (https://datausa.io)

Gapminder   Gapminder is a large collection of data sources from around the world including the World Bank, the United Nations, World Health Organization, International Labour Organization and Forbes among others. Data can often be regionalized and is fully downloadable. (https://www.gapminder.org/data)

Google Trends   Google Trends is a powerful tool that enables you to look at searches over time in order to see if certain words or phrases are increasing in popular search usage. Beyond this, Google Tools can be used to ask

FIGURE 4.6 Final score

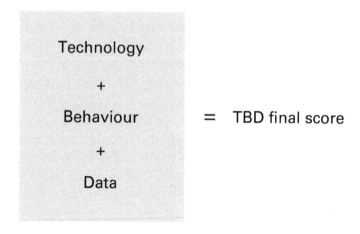

very specific questions in order to gain in-depth analysis based on geography and related searches. (https://www.google.com/trends/explore)

Step 4: Calculate your TBD score

In order to find out the final TBD score for your question you need to add up the three scores from the Technology, Behaviour and Data sections.

Remember, if you are using the recommended 0–10 scale, the maximum score for TBD can only ever be 30. There is never any subtracting, multiplying or dividing involved for simple TBD.

TOP TIP

From experience, the score of 1–10 gives enough room to manoeuvre and is universally understood but you may want to alter this or use 1–100, traffic lights or another system – whatever works best for your company – but above all, it must remain consistent over time and remain equally weighted.

Now you have your final score, it is time to find out what decision you will apply to the issue or question you posed using the Decision Matrix you created at the beginning.

Step 5: Apply TBD score to your Decision Matrix

Now you have your TBD score it is time to assign a decision to that score. As previously mentioned, everyone's Decision Matrix is different but it is important to remain true to whatever decision you have agreed to. Clients have told me it is tempting to knock points up and down based on the time of day, energy levels or current workload at that time. I urge you not to do this for a few reasons but mainly because you are not being true to the aims of the process. Everyone gets busy and earlier chapters talk about this as it is an important part of making the TBD process successful.

Identify what the score you achieved corresponds to and write it on the top of a piece of paper (or email or whatever medium you work best in). Now start to flesh out the decision based on your understanding. The best advice here is to not put it off – do it then and there. This way you'll have everything fresh in your head and get the best first thoughts down. You can of course return to it later, add to it and have others contribute – the key is to keep the momentum up.

You've taken the time to get this far, so don't sabotage yourself – stay true to the decisions you made. Don't make yourself feel better in the short term to make yourself feel miserable in the long term.

CASE STUDY

Before we look at the Advanced TBD framework in the next chapter, let's look at an example of simple TBD from a client case study.

Susan is a CMO of a well-known fashion retail outlet, which is considering implementing a new type of changing room. The new changing room is drastically different to the old style of taking things in and leaving the customer to it.

The new changing rooms:

- replace the cashier;
- allow customers to browse more clothes than are in the store;
- augment the general experience to be more fun and engaging;
- allow people to try items on without actually trying them on, using a projector;
- enable customers to ask advice from online stylists;
- allow customers to pay in the changing room and organize delivery (if they don't want to carry it around with them).

1   Technology (CAN?). While everyone can shop like this, not everyone goes to physical stores to buy clothes so it cannot be a perfect 10 [SCORE = 8].

2   Behaviour (WILL?). Clothes purchase is a personal thing to every individual. Some take time and effort when selecting clothes whereas for others it is a quick in and out purchase choice. Others purchase online and this is an increasing trend. Therefore, the score reflects a majority supporting the idea but not a higher score [SCORE = 7].

3   Data (ENOUGH?). Based on similar new-style changing room examples, people are responsive to additional information while trying clothes on. The CMO found some interesting additional questions to ask around body scanners in changing rooms and also identified that costs are lower when reducing staff, distribution and inventory held on the premises. Additionally, the CMO found data about increased basket size when clothes fitted better (people bought two versions) and customer satisfaction increased when clothes were delivered to their home rather than having them with them as they walk round other shops [SCORE = 8].

Total TBD score: 8 + 7 + 8 = 23

Based on the Decision Matrix Susan created with her team beforehand, the score falls into the company's middle category, 'Create a small test case', which falls between 'Research further' and 'Implement widely immediately'. The score means there is automatic budget assigned to fund a test (more can be applied for but this will require a pre-agreed higher-up sign-off, although this test is mainly about rerouting existing functionality). Based on this, Susan determines that a one-store test is most appropriate and activates a two-person team to own the project (with her oversight) to be completed within two weeks and then implemented within a maximum of one to two months.

Conclusion

This chapter gives you an easy method to get things moving and lots of resources to enable you to make better decisions. The chapter also helps other areas of the business understand the rigour you are attaching to making decisions and how you think about disruptive technologies.

TBD can be a simple process but there are limits to when it can be used without additional information and interrogation being needed. Subsequent chapters will deal with these issues in addition to creating open businesses (and departments), things to watch out for to maximize your chance of success, and how to gain sign-off (including funding) at all stages along the way.

The following chapter moves from simple TBD to the more complex framework, enabling you to prepare the necessary elements required to complete your own advanced TBD framework.

05

Complex TBD

The coming chapter will enable you, your business or department to:

- follow steps to complete your own complex TBD framework;
- identify core areas of interest and areas that can be de-prioritized based on your own company objectives;
- avoid common pitfalls when thinking about disruptive technologies and emerging technologies;
- maximize the chance of success and secure additional funding if/as required.

The previous chapter focused more on a light touch (or use) of the larger TBD framework. In this chapter, we will explore an advanced version of TBD or (TBD+) which will enable you to create a focused grid of investment – be that money, time or attention for your business or brand. In this chapter, you will learn how to prepare the necessary elements required to complete your own advanced TBD framework.

Why are two versions of TBD needed?

As you saw in Chapter 5, TBD was born out of a long conversation and was always meant to be a quick decision tool. As time progressed and the framework was used and refined, additional elements were added to increase its usefulness and robustness, include other people and above all enable it to be embedded into an organization rather than to be a one-use item. Naturally, if your company feels like a more rigid system wouldn't work (or just has a reactionary style), try just using simple TBD. Simple TBD wasn't meant to be a long-term solution; TBD+ is. Companies often require specific, longer-term views of emerging and disruptive technologies (an often difficult and laborious task) rather than simply creating or receiving a list or a timeline – TBD+ was created to streamline this process.

What is TBD+?

Unlike simple TBD, TBD+ brings in the individual company's priorities and areas – after all, it is somewhat futile to follow something like nanotechnology if it has been deemed there is little to no likelihood of it impacting your company. TBD+ creates an editable grid of technologies and identifies key areas that your company can excel in based on a system that includes multiple criteria that are unique to each company.

In short, the TBD+ process allows anyone to create a prioritized grid of technologies that enables a company to focus its resources and make the correct investments, whether that be time, money or both.

When do you use TBD+?

Unlike simple TBD, TBD+ is a tool that takes a little more time to create and is meant to be updated on a quarterly basis. Both these features are key differentiators from simple TBD but the initial setup is usually the part that takes the longest. Once you have set up the various elements, the updating and checking is usually pretty simple. Clients vary depending on their individual levels of comfort with the system and different circumstances but most clients (after the initial creation) tend to update TBD+ on a quarterly or bi-annual basis. By doing the quarterly updates, companies ensure the process is not just bedded in but is also a guiding star and helpful reminder for subsequent strategic decisions that may need to be made.

TOP TIP

Use your calendar to get the most out of TBD+. Setting recurring reminders and notices in your calendar right at the beginning of the process is a smart move that often makes the rest of the process easier. From reminding the group to identify new technologies to collecting data yourself by putting in monthly, quarterly and other big milestones, the TBD+ process becomes more ingrained in the corporate space. I have found it incredibly useful to set aside time to just simply plan and think big before meetings. For more information regarding clawing back time refer to Chapters 2 and 3.

Before you start, a word on failure

Previous chapters of this book and ones to come will discuss the difficulties of organizational change. You will have challenges, challengers and bumps along the way. Give yourself, right now, permission to fail. Not the whole process but parts and times along the way. Free yourself from the pressure of perfection. Do it now. Say it out loud or write it down but do it. Nobody is perfect in business (or in life) and without the freedom to fail people rarely reach for anything too far. You aren't one of these people.

Now this doesn't mean that what you produce cannot or doesn't have to be perfect but the process or way you get there is unlikely to be smooth or simple. Everyone hates to fail but the best way to get over failure is to expect it and then when it comes along you are less likely to be completely bowled over by it. There have been many quotes pertaining to failure that stick in my mind but one I use regularly with clients when things go awry is adapted from a Thomas A Edison quote: 'Failure is just another way not to do something. Find the way that does work.'

With a 'confident failure mindset', as I call it, you create the expectation of failure. In other words, you create a realistic perspective – when was the last time you (or someone you know) got something complex and strategic totally perfect right off the bat? By bringing up failure upfront, you introduce a refreshing pattern interrupt that makes people sit up and think – what's happening here? I don't have a pre-programmed response for this so I had better hear more. Again, you aren't focused on the failure of the entire plan, just that the route to get to the end may not be the one you set out with.

Now you have the why, the right mindset, the focus and the desire, let's start going through the steps of the TBD+ process.

The TBD+ process

Step 1: Define your company's objectives

Having worked with a wide variety of companies – big, small, start-up and global brands – one thing continues to amaze me when I speak to employees at different levels. Few of them have a clear vision of the company's objectives. What is worse, many return responses that are often negative towards senior individuals, inferring that there are already divides within the company that will make my job harder. Few companies have clear visions and objectives beyond perhaps a quarterly update or a yearly speech. Often what's worse is that taglines become objectives.

The point of this is to labour the idea that visions and objectives are crucial when it comes to the future. These seemingly pointless statements and taglines do add up to something, both internally and externally. If people aren't 'singing from the same hymn sheet' you are disadvantaging yourself in a time when you don't need more challenges. If you cannot write down two or three objectives for what your company (not your role or the quarter you are in) does and what it is looking to achieve, start researching.

How to do this: often with these companies, such information does exist but it isn't often discussed, updated, challenged or promoted for a variety of reasons. In this instance, I often ask who created the information, when was it created, is it still true and can it be adapted?

Places where you can find the information include CEOs, Investor Relations areas on websites, press rooms and sites like Wikipedia and corporate sponsorship blurbs. Once you have these objectives, lay them out in front of you on separate bits of paper and think about how these filter down your organization and what they mean to you. This may spark some questions for senior executives immediately (like the ones above) – write these down and make sure you ask them before proceeding.

Once the list of objectives is finalized, write them at the top of a sheet of paper. These are your guiding principles. The reason you are embarking on TBD+ is to create a robust strategy for the future – understanding your guiding principles gives you a focus for the whole process. Some will be extremely relevant to the journey while others will be less so. The point is, you have a pointer to the horizon. Don't forget there are numerous other factors that will determine the success of the TBD+ process and your company's future. You are using TBD+ to chart a number of technologies and fields that will help you navigate towards your goals. Commit to the process and judge progress accordingly – every day you are a step closer to the goal (which will then shift once you reach it).

Step 2: Assemble your team

A steering group (sometimes called a 'steering committee') usually decides the focus of a business or organization and how it is run. The right people are not necessarily just ones you get along with or have worked with on successful projects in the past – these are strategic and key players within the business.

While it could be argued that involving more people complicates things, experience tells me TBD+ provides the best results when others from the organization are included:

- The process being subjective and having multiple viewpoints enables you to come up with a number that is more representative of where your customers and business are.

- By involving more people from the organization the process and, more importantly, the results are more likely to filter through the organization more efficiently.

- Involving people with different viewpoints and experiences means potential kinks, issues and roadblocks can be seen, avoided and navigated before they become problems.

- When people aren't involved in projects it is often easier for them to criticize and derail things after the work is done.

- Each person will be specialized in different areas and so different viewpoints and opinions can be obtained quickly.

- More people involved means that the likelihood of failure is lowered as there are more eyes and hands to carry the burden.

While it is absolutely possible for an individual to complete TBD+ it is not recommended. Experience has shown that this not the best route as the specialization of each of the recommended members (and their cooperation) makes the process more robust and less prone to individual biases. Increasing the amount of people involved can also decrease the stress and pressure on the individual conducting TBD+.

TOP TIP

If you do decide to complete TBD+ on your own, take more time on each section and make sure not to skimp on the earlier steps as these are vital to a strong TBD+ score.

YOUR TEAM

The process will require the following people – naturally job titles and functions differ from business to business but generally the following people are required:

- TC (technical consumer) – Head of Brand, Marketing, Sales, Communication.

- TO (technical organization) – CTO, IT Manager.

- BC (behaviour consumer) – Head of Customer Service/Customer Experience.

- BO (behaviour organization) – CEO/Managing Director.

- DC (data consumer) – Chief Innovation Officer, Consultant, third-party adviser (e.g. consultancies like Bain, Forrester) or your agency of record (if appropriate).

- DO (data organization) – Chief Finance Officer, Head of Commercial, Management Accountant (whoever is responsible for the current and future financial health of the organization).

GETTING PEOPLE INVOLVED

One of the best ways I have found to get people excited – or at least understand why they are being tapped to be involved – is to send them a killer stat or a really thought-provoking quote from an article from a respected title.

A lot of people will be uninterested or won't understand why they are being tapped for this activity. Once you explain the process to these individuals and discuss why it is important to the future of the company, a lot of this goes away – some may still resent being involved as it is out of their job description but try to bring these people on board through public (show of hands, voting mechanics) or private means (1-2-1 explanations, help, advice). It is important to note from the outset that just because their name will be going on the team list, it does not mean they cannot get other people involved – in fact, some organizations produce better results when this behaviour is encouraged.

Doing anything new and untested within an organization is risky. Some work best when things come from the top down and others work best when innovation and change come from the bottom up. You can determine which way your company works – crudely – by listing all the major news and decisions that have come from your business over the last six to twelve months. Now mark them either top down (if the senior management team decided them) or bottom up (if those initiatives came from the brains of the entire workforce). Neither is right or wrong but both have an impact on the way a business operates and communicates. The next chapter discusses selling change and the process in more detail for either business type.

THE FIRST MEETING OF THE GROUP

Before the group meets for the first time, decide carefully on the messaging, other attendees (you may or may not want your CEO present) and the location. Some organizations prefer to conduct the meeting (and TBD+) in-house while others may want to add a little theatre or breathing room to the proceedings. The best recommendation is whatever you think will help people understand and come together as the unit you need them to be. It is important for this group to identify with each other and understand the significance of what they are being asked to do. When the group meets for the first time, talk openly about the time requirements, the importance of being thorough, the help they have (and can ask for) and the 'why' behind the tasks. Beyond this, the usual important meeting recommendations apply:

- Make sure the room has lots of natural light.
- Make sure it is well ventilated.
- Make sure there is plenty of time to ask questions.
- Make sure there are enough comfortable seats.
- Make sure people know when to show up.
- Do not have the meeting unless all required people are present.

TOP TIP

Using words like 'start', 'give', 'we', 'want', 'choose', 'move' and other positive words has been shown in studies to produce more positive group cohesion. Beyond simply changing the words you use, try to empower others and help them see the process as one that isn't a nice-to-have exercise but one that will determine the path of the company in the next one to three years (it's fine if it goes beyond this but be realistic with staff turnover, current mood of the company and other factors). The point of this timeframe is to make it a short enough period of time for them to have an impact or be impacted upon.

Step 3: Define your target group (or groups)

Modern businesses operate in highly competitive, open and global markets (often even if they are only in one territory) because of the internet. As we have seen in the first few chapters of the book, this world is in flux. The people living in it are also changing because of this and other factors.

Consumers today are more demanding than ever before, more connected than ever before and have more options than ever before. A key area of TBD+ is to understand your target group better than your competitors do. Before reading on, stop for a moment and read that last sentence again – can you say, right now with your hand on heart, that this is true for your organization? Do you know your target group better than your closest competitor does? What about the challenger in your sector? What do they know that you don't?

People today are demanding, complex and often contradictory beasts:

- They demand more of everything for less.
- They expect privacy but give out personal information to strangers.
- They have more time but claim to be busier than ever.

The only problem with these statements is that they are broad, sweeping generalizations; you could not apply them to both a 20-year-old in the United States and a 60-year-old consumer in Hong Kong, for example. Beyond age differences, the social and geographic differences are too vast. Knowing the intricate behaviours, beliefs and patterns of your specific demographic is imperative to the TBD+ process. Your first job as a team is to agree and define your business target group (or groups) in order to make your TBD+ as specific to your business as possible.

Being specific at this stage is critical in focusing in on key differences, areas of similarities and point of interest for the future. There are several reasons for breaking down your business into multiple target groups but the main one is because stereotypes or age ranges aren't enough – graduates have different specialities, parents can be young or old, people can be introverts in their home lives but extroverts when using your product. Making sure to spend the time really lasering in on who it is you are targeting (and who you aren't targeting right away) will save you time, money and effort in the long run.

Step 4: Identify a clear technological and behavioural profile of your target group (or groups)

Once you have your company objectives, have assembled your team and defined your target consumer group, you need to flesh out what you are up against. This can be done in multiple ways including hiring an advisory/ consulting firm like HERE/FORTH to come in and present to you. Making sure you understand your target demographic beyond their purchase cycle is key in moving markets and disruptive times. Having a dynamic and crystal-clear picture of every aspect of your target's life means any surprises (or disruptions) are less impactful, can be taken on board more fluidly and will enable you to spot more opportunities which could potentially be lucrative or create additional revenue for the business.

Below are some tools that, while not exhaustive, will give you a good steer on exactly what your target groups are doing and thinking.

Google Analytics   This is possibly the most important source of information for you (or whatever internal analytics service you use) as it will tell you what is actually happening when people get to your website. Whether you sell through your website or it is simply an information-based portal, knowing the routes consumer groups navigate around and find your site through is integral for this step. (https://analytics.google.com)

Canvas8   Understanding the why behind the behaviour is often more important than the insight itself – this is where Canvas8 specialize. Through their extensive library of insights and the knowledge network they have built over the years (including an impressive reactive model), the team produce bespoke reports for clients that increase understanding of different behaviours and demographics. (http://www.canvas8.com)

TGI   Owned by Kantar Media, TGI is a smart tool that enables users to find out the likelihood of a target group using other technologies, among other areas. A powerful database of regularly updated information for cultures around the globe, TGI can be used to analyse and test hypotheses about different demographics using thousands of variables. (https://www.kantar.com/expertise/advertising-media-pr/consumer-profiling-and-targeting/tgi-consumer-data)

Ipsos MORI   Based in Europe, Ipsos MORI works globally with big brands to help them understand multiple aspects of their business including brand communication, advertising and media research (Ipsos MORI Connect), consumer, retail, shopper and healthcare research (Ipsos MORI Marketing), customer and employee relationship management research (Ipsos MORI Loyalty), and social, political and reputation research (Ipsos MORI Public Affairs). Lots of free data is available but bespoke information and research is where the smart money goes. (https://www.ipsos-mori.com)

Mosaic   Created by Experian, UK-focused Mosaic is a 'cross-channel clas-sification system' which means they create consumer profiles based on a wide range of data. Containing a lot of data points, the database can be segmented and cross-tabulated to create bespoke profiles and enable you to understand your consumers in extraordinary detail. (https://www.experian. com/marketing-services/consumer-segmentation)

Forrester Consumer Technographics   These are dynamic dashboards that enable you to understand your consumers' purchase paths and behavioural traits in addition to helping guide your strategy and identify untapped consumer groups. Expensive but worth your time. (https://go.forrester.com/ data/consumer-technographics)

Google Consumer Barometer   According to Google, 'The Consumer Barometer is a tool to help you understand how people use the Internet across the world.' Using graphs, comparison charts and other interactive elements, this free-to-use service runs in association with multiple industry names and can really help flesh out customer journeys. (https://www. consumerbarometer.com/en)

Emplifi   With a raft of free data, reports and insights, Emplifi (previously Socialbakers) are one of the busiest social media analytics firms out there. Beyond the tools and services that help you understand your existing audi-ences on platforms like Facebook, TikTok and Twitter, Emplifi can also offer brands insights into competitors' communities – handy and vital to fully understanding your target group. (https://emplifi.io/)

GWI.com   Beyond the simple (yet robust) datasets available, GlobalWeb-Index's database and research panels around the world enable you to create custom research queries and interrogate the granular information GWI have in order to really get under the skin of your consumers' digital footprints. See Chapter 4 for more details. (https://www.gwi.com)

MRI-Simmons   Boasts the 'largest and most current database of consumer behaviour, media usage and consumer motivations' and is a serious source of information for any brand. Predominantly focused in the United States, some insights and data points can of course be used outside this territory. The Survey of the American Consumer® is a particularly worthwhile docu-ment to keep up with although MRI-Simmons offers an array of specific services from magazine to viewer research, consumer segmentation and cross-media analysis. (http://www.mrisimmons.com)

VisualDNA   Founded more than a decade ago, VisualDNA combined several scientific approaches (data science, psychology, engineering among others) and fused them with creative undertones, creating a unique product to help brands understand their consumer. Via interactive and virally spread quizzes, VisualDNA enables brands to really understand their target consumer (and importantly their surrounding communities) in a unique way so they can fully understand the motivations, interests and personalities of any target group. (https://www.visualdna.com/profiling)

TOP TIP

Use Facebook Ads to create a small, fast temperature check on your brand, and your community's likelihood to do (or not do) something. We used Facebook Ads containing a questionnaire for clients who needed to understand their Facebook community better. Utilizing the paid-for ad platform that Facebook has, you can target your own community and serve ads with a link to a questionnaire site like Survey Monkey or use a Google Form. Ultimately, you can ask any questions you want but this method is best used when quickly deployed. Prizes and incentives aren't usually necessary but you may want to collect and use the data so some sort of reward may well be warranted or prudent.

If you are looking for more insights from your own data, Adobe Marketing Cloud, Oracle Marketing Cloud, Umbel (big data and graphical visualization) and AgilOne (predictive analytics) are all great providers of data analytics and insights for business of all sizes. Additionally, think about specialist companies who could provide you with wider information too like trends (Foresight Factory, Trendwatching, The Future Laboratory, Protein), UI/UX (Webcredible, Punchcut, Rossul, Tuitive Group) and eye-tracking (Tobii).

Once you have gathered and assessed your data from the various sources you choose, it is important to share your findings and agree them as a group. Below are two exercises that HERE/FORTH uses with clients when helping them go through the TBD+ process.

1 SCRAPBOOKING WORKSHOP

This is a simple workshop that allows participants to demonstrate knowledge and propose questions to the wider group.

You will need:

- tables (round tables are best);
- magazines from a wide variety of subjects;
- additional media (leaflets, printouts of websites, books, logos);
- scissors, glue, blank paper, markers, rulers, erasers, whiteboards, easels;
- plenty of space;
- computers and printers (optional).

Time required: 45–60 minutes.

TOP TIP

Leave plenty of space between tables and groups; make sure people stand up and move around rather than sit down as this often facilitates more energy and ideas.

Break the group up into smaller groups of five to seven and ask them to create a visual depiction of your target customer using images, words and anything else they can create using the tools you have provided. Make sure all participants are clear on what they are expected to produce and the desired outcome. This can be done for the same target group or different ones but there should be some overlap – i.e. each target group is covered by more than one team. You are looking for a full picture of your target demographic; what they eat, use, like, dislike, what brands represent them and so on. Instruct your group to go as deep as possible into their knowledge and if they can't find the image they are looking for just to write it down instead.

In addition, ask the teams to create three questions about their target group they don't know the answer to. The questions could be anything from 'What do they feel about 3D printing?' to 'What time do they go to bed on the weekend?' Collect these answers anonymously if you think that will help your group yield better results. When these questions are collated, look for crossover and get the group to openly discuss them and how to answer them. If this process will require more time, make sure to feed back to the group and then follow up with answers via email and in person. While prizes are not required, some sort of trophy is often a fun way to encourage helpful competition and focus people.

The end result will be a diverse range of images and 'mood boards' that depict the same target. As a group, discuss the results and create a single board that is a combination of the target group to use as a starting point moving forward.

Ask open-ended questions like:

- What surprised you in this board?
- What are the key points? Why?
- What do you think could be added to make this a fairer picture?
- Is there anything you wouldn't have expected to see a year ago?
- How do you think this might change in a year's time? What about five years?
- What do you think this board says about how we should talk to them?

The more questions, the better – keep the conversation going and keep asking people why they think what they think in order to get deeper insights and find out why people believe things to be that way. This can be uncomfortable for some people so make sure to be reassuring and include everyone.

If you aren't sure about an element, challenge it constructively and either agree to change it or find some evidence to solidify the objection before confirming it. At the end of the session you will have a mostly finalized version of your target demographic. This image should then be tidied up and put into a format that can be distributed around. You will also have a list of questions that need answering in order to cement the correct information about your target group in the minds of the workshop attendees. The workshop is worth doing at least once a year or when a new product is set to come out.

2 CUSTOMER (OR JOURNEY) MAPPING

Customer mapping or customer journey mapping is the process of creating a graphical vision or representation of the experience your customer has with your company. This task is a complex version of the above that takes into account different data points in order to create a clear map of the consumer or target group for your company.

You will need:

- data (internal and external);
- Post-its;
- whiteboards;
- plenty of space.

Time required: 45–60 minutes.

This group exercise is excellent at highlighting the relationship or relationships between a brand and the customer beyond the initial purchase or spark of interest. The point of the task is to identify areas of weakness, opportunities and potential areas of consumer loss that can be plugged and used to identify areas of investment. Customer journey mapping can also be used to identify new areas of investment as new needs, experiences and opportunities are also often identified during the process.

TOP TIP

Both of these exercises can be completed by a single entity but the best results usually come when a diverse team is pulled together. You may want to bring in more people from the organization (including marketing, IT, SEO specialists and other experts) to help you form the best map possible for this exercise in particular.

The first step of any mapping process is the collecting of data. In the case of target consumers, start with asking the groups (4–5 people per group is best) to shotgun down sources of information that could be used to gain observations about customers and their experiences. Ideally, team members will be from across the business and a wide range of behaviours, insights and information will come out. Key individuals to include in this session are people like the CTO, CIO, CFO, Sales and Customer Service – each of these departments has key insights into the same people; understanding how they all impact and fit together is key to unravelling how your customers go through the process of buying what you are selling.

Use your own data as a starting block. Asking website developers and SEO experts and looking at media spends are crucial to understanding how customers find you. Use this as a starting block but expand out as these are only existing customers. The customer journey mapping exercise is also about potential customers in your target demographic.

You may want to put this data into three categories:

1 what we know;

2 what we think we know;

3 what we don't know or don't know we don't know.

TOP TIP

Test your own assumptions. Often people hold on to old insights, data, nuggets of information and outdated thinking without knowing they are doing so. A tip for making sure your data is fresh and the best it can be is to ask yourself after every insight or tip is shared, 'How do I/you know that?' Asking this question helps sift out what is fact, what is interpretation and what is misunderstood as fact; it can save costly mistakes now and down the line. Each category needs to be carefully evaluated and weighed accordingly to make sure you aren't missing anything important or aren't necessarily biasing the process.

The second step is to collect these insights into a single place and share the findings with the entire group to unify thinking and make sure everyone understands the target customer or target group. Often this is done in a vision board style but it is useful to also take some time and bring the customer to life through vox pops, bringing in the technology the person uses and showing percentages etc. A client once turned an office into a room representative of their customer profile in order to help employees visualize a day in the life of their target consumer.

The third step is to go beyond your own data and think about what else your target demographic is likely to be doing at specific times throughout the day. This layer of information adds important context to your map so do not skimp on it. Look at external reports, conduct customer surveys; qualitative and quantitative studies can be invaluable at this stage. Understanding the why behind these activities is key – companies like Canvas8 can be used to really understand the psychology and mentality of consumer groups' purchasing behaviour and attitudes towards other aspects of their lives.

Once you have collected all the data on your target group and have assessed it for validity and applicability it is time to arrange it and visualize it in a way that will enable you to see gaps, opportunities and areas for improvement.

The best way to do this is with a large area (like a wall) that has time across the top of it and then use the data you have gathered to fill in a day in the life of your chosen target group (or individual within the group). Think about the life of that group or person from the second they wake up to right before they go to bed. This process takes a while and should be as evidence-based as possible. Think about needs, perceptions and processes associated with your company and purchase journey.

Some thought-starters for you:

- What do they do right before they go to bed?
- What sort of alarm system do they use?
- Do they get a paper delivered?
- Are they a member of Amazon Prime?
- What type of smartphone do they use?
- How many hours of sleep do they get? Why?
- What apps do they use every day? Once in a while?
- How did they find out about your business?
- Are they a member of any loyalty schemes? What does this tell you about them?
- What products do they buy regularly? How do they buy them?
- What sort of computer access do they have?
- What do their living arrangements look like?
- What relationships do they have?

There is no correct way of visualizing the final result – some prefer a linear look at their customer, others prefer stories, while some use this as a stepping-off point for greater research. Lego produced an impressive take on this with their 'LegoWheel' (experiencematters.files.wordpress. com/2009/03/legowheel.png), which shows how their brand can impact a customer before, during and after they fly to Legoland. A nice touch was the 'Experience Icons' that Lego used to denote points of 'happiness', 'data' and 'make or break moments' in order to decide what role they could and should play at different points along the way. A very mature way of looking at the customer relationship.

This exercise enables companies to have a formal approach to describing, designing and understanding experiences for customers – not just purchase paths. Thinking about who owns the target customer's attention at different times and in different areas is key to understanding how, when and if you should disrupt or impact the journey. The simplicity of the process is underpinned by the data that it uses and the commitment when creating it. However, it does not live in a vacuum. The findings of this exercise must be shared company-wide in order to make sure insights and opportunities are shared – not just quick wins.

In later chapters, we will discuss taking things like this forward and successfully getting senior buy-in and higher-up approval on things that may have a somewhat unknown outcome.

> TOP TIP
>
> In order to keep the maps from being outdated and showing inaccurate reflections (what you have produced is only a snapshot), you should create a small, agile team to identify and update the map using specific indicators that the group determine are most likely to change over time. These could include customer feedback, R&D and technology profiles among other options.

THERE IS A SHORTCUT...

Both of these exercises and others like them, as well as working with different agencies, enabled me to create the Consumer Demographic Canvas (find at www.hereforth.com). This is a one-sheet exercise that enables a company to quickly spec out their consumer to identify areas of weakness and strength. While the sheet is a useful capture document it shouldn't detract from conducting the workshops, as it is only through these that subtle nuances and specific data can be identified.

> TOP TIP
>
> Take some time to construct your own Consumer Demographic Canvas that is specific to your needs and industry. Brainstorm with industry leaders and people inside and outside your company. Make sure you go beyond your department, think big and get all the data you can – you might not use it all but it is good to know what you don't need.

Step 4 is one of the longest areas of TBD+ and while every company I have worked with is different, this part averages out at around 5–10 hours in total.

Step 5: Create your Investment Matrix

The Investment Matrix is similar to the Decision Matrix in Chapter 4 in the sense that bands are decided on and then implemented but the difference comes in the outcome. Instead of multiple types of decisions that can be made (further research, do nothing), there is only one type of decision (time, money or neither) with multiple levels of investment agreed on after careful discussion. Some companies I have worked with have decided to implement smaller investment plans and both options are valid and useful for different businesses. In order to keep things simple, we will discuss how to implement the former here.

As with the Decision Matrix, there is a maximum score, although as there are now six scores, the maximum that can be assigned to any technology is now 60. The first step is to create the investment scenario, outcome or decision you will make should a technology be given the 'perfect' 10 score by all participants (or on all axes if you are running TBD+ by yourself). Once you have this score, the next step is to look at the other end of the spectrum and determine what a zero score would return. Just because a technology returns a zero score does not mean that it won't be useful or become of interest in the future so think hard before you dismiss it out of hand.

Again, as we saw in the Decision Matrix, wishy-washy answers are not allowed in the Investment Matrix. The process for determining the different bands or specific numbers is the same but the outcomes of each must be specific. These can and will be different for companies and industries in the same sectors but the point is to make it unique to your company. Before we dive in, remember some key pointers from Chapter 4 or perhaps refresh yourself with the entire process.

Key points:

- TBD is for quick decisions, TBD+ is for investment.
- Make sure the outcome is specific, action-oriented and realistic.
- Take time to decide the different bands of the Investment Matrix – investment is a larger decision than simply researching into something.

Examples of some of the investment best practices that clients have come up with in the past include:

- Within 12 months of agreeing final investment requirements, [COMPANY NAME] agrees to design, run and evaluate a [SCALE] test pilot for [TECHNOLOGY]. If successful after a [TIME] period, the test will be rolled out further. The level/scale of this will be determined by [PERSON 1, 2, 3] of [COMPANY/DEPARTMENT].

- Further time investment is required at this stage to determine the viability of a [BUSINESS TEST PROCESS] for [TECHNOLOGY]. Therefore, within 72 hours, a further round of research will be conducted by [PERSON 1, 2, 3] in order to clarify the following issues for [COMPANY]. If this is not determined within 72 hours, a further extension may be requested (but may not be granted). After the deadline has been reached, a final TBD+ score will be created and a decision will be finalized by [PERSON 1, 2, 3]

- Action is required. If the cost of [BUSINESS IMPLEMENTATION] is less than [AMOUNT] you are authorized to proceed and report back to [TEAM/INDIVIDUAL] every [TIME PERIOD]. If over this amount, a project leader must be assigned who must then fill out the short business case with the funds request for initial approval.

These will be different for every company; the idea behind getting these written down beforehand is similar to the reasoning behind the Decision Matrix:

- doing so provides accountability;
- people are aware of the potential outcomes available;
- people have had a say in creating them so are less likely to fight them;
- if something is written down it is more likely to be achieved.

Once you have your best and worst outcomes, decide on the other numbers, zones or bands you want to have on your Investment Matrix – remember there is no right answer. Some businesses want lots of outcomes, others only want a traffic light system. The best advice is to make sure you have enough outcomes to satisfy your corporate culture and help you make easy, practical first steps.

Remember: the most successful companies that use TBD created fewer bands for decisions (i.e. with five or seven points between bands) with specific simple actions for the lower bands and intricate instructions for upper bands

FIGURE 5.1 Investment Matrix

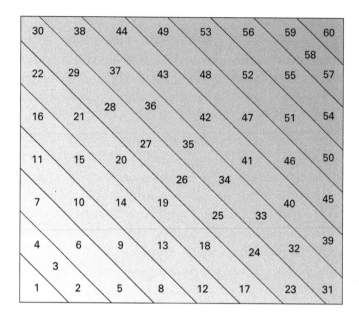

rather than lots of complex decisions for each number or lots of numbers. Fewer bands enables faster movement through the process but often means questions crop up afterwards. Make sure you decide early on what works best for the people involved in the process and those who need to sign off on the outcomes it creates.

As with the Decision Matrix, more hands make light work (although it may not seem like it at the time) so get other people's thoughts before the bands are locked. Remember to talk openly about best cases, likely cases and how the company has dealt with similar things in the past. Sometimes, because sums of money are involved in TBD+, people are more cautious with assigning higher numbers to technologies – make sure to think about this when creating your bands and discussing your outcomes.

Step 6: Identify existing areas of technological interest

The next step is to identify key areas of existing technology investment and interest in the Investment Matrix. These must be areas that are already of interest to your business – nothing in the future at this stage as these will need to be put through the TBD+ process. Naturally, these will be different from business to business (and possibly within the same industry) but think about the whole process undertaken by your business and the consumer. Think about what is business-critical and what is a fringe technology you may use but perhaps don't rely on heavily – both need to go in but they won't go in the same areas of the Investment Matrix. To give an example, think about a fashion retailer; they would use point-of-sale technology (critical to their business), they use in-store beacons to offer personalized coupons to shoppers (not critical but useful) and they might use automatic check-out stations (not critical but they definitely help the business get money). Each of these would go in various areas of the Investment Matrix based on the money you have spent (not that you plan to spend) thus far.

When creating this list – and it may be as long or as short as you like (it can always be added to) – think about the following areas for inspiration:

- existing platforms you are already engaged with;
- existing technology you use;
- existing partners you work with.

On the list you might include social networks (Facebook, Instagram, Flickr), mobile messengers (Snapchat, Slack, Workplace by Meta, Whatsapp, iMessage), search companies (Google, Microsoft), advertising companies (Yahoo!, Facebook), location data companies (Yext), and marketing technologies (contactless payment, live research companies).

The key is to think about the top level of the technology you are considering – instead of TikTok or Facebook Live, think general live-streaming video. This will enable you to see more opportunities for technologies that emerge and potentially disrupt these players rather than simply having both their names in the grid.

Remember: there are no wrong answers to this question – you can prioritize and sift out less important technologies later on.

Step 7: Plot these on the Investment Matrix

Once you have the list, the next step is to place these existing technologies on the Investment Grid. The simplest way of doing this based on working with large and small companies is by seeing all the technologies laid out in front of you and then moving them around in order of importance. The reason we aren't putting more rigorous formulas or calculations on these at this time is that the technologies are already in place and you need to know where they are now rather than where they could or might be in the future. Use this exercise to look at the business from the top level – if anything jumps out at you (like being weak in robotics, for example), write this down for use later.

TOP TIP

Using Post-it notes or pieces of paper on a makeshift clothesline is a great way to make this task more interactive and really compare different items in relation to each other.

Step 8: Identify potential new areas of technological interest

This is often the hardest part of TBD+ as you may not know straightaway what the areas of interest are.

Think about trends you've heard about, conference topics you've seen, use Google Analytics to get an idea of other words associated with the technologies you already use, look at analyst reports (and blog posts – these are sometimes more frequently updated), look for area or topic curators on platforms like Twitter and LinkedIn, ask a question using a Q&A site like Quora (quora.com) or find an expert in the field and take them to lunch to discuss areas that cross-connect or collide with the one you are interested in.

THE EXPERT REFERRAL STAIRCASE

This system was learnt from a mentor when I lived in Los Angeles at the start of my career. I didn't know a lot about the lie of the land and he taught me a way of finding out top information in a robust (but understandably not foolproof) way.

The system works like this: you start out having an idea of who may be able to help you understand something a bit more (let's call them 'Expert 1'). You chat with Expert 1 and get the information you can from them but then the key here is to ask one simple question at the end of the meeting. This question can change but the idea is the same – you are looking for 'their' expert.

The question may be:

- 'Who do you turn to for information and trends in this area?'
- 'Who is at the top of this field in your opinion?'
- 'Who would you go to if you had a question about this?'

This next person or source is Expert 2; once you have this information, you go after them and the cycle repeats until you are satisfied you have the best answer or until you find another topic in order to go into another area. The idea is that every time you question someone you go up a stair and get better and better information. You constantly learn new information throughout the process and often find new areas of information to explore. Beyond this, you also create for yourself a panel or collection of smart people you could also potentially cultivate into a think tank or some sort of knowledge collective for the business.

TOP TIPS

Before you begin this questioning process, make sure you have written down clearly what you are looking to achieve – lose all the extraneous information and focus on the deep issue you are looking to find out about. Are you, for example, looking for revolutionary solutions or do you need practical ones? Another tip for this method is to not just look for academically smart people – look for the people you know (or that the expert knows) who are creative thinkers. Finally, keep the process light and, ideally, free from prior knowledge or other constraints so that you get the experts' best thinking rather than their first idea or just any solution that they think will work (or that you want to hear). The process takes time and is an art.

There are lots of other ways to find this information internally and externally – the best way to do this is to be open, hold concise brainstorms and do lots and lots of research and mapping.

The good news is that you already have five technologies to put into the grid:

- nanotechnology;
- 3D printing;
- blockchain/bitcoin;
- artificial intelligence (and machine learning);
- holography.

Other examples could include virtual reality, augmented reality, chatbots, point-of-sale technologies, Uber, Netflix, the sharing economy, Meta, autonomous vehicles, robotics, home-testing genomics, drone surveillance, wearable technology, decarbonization tech, on-demand drug manufacture, wireless energy, self-fertilizing crops, edge computing, data fabric, solar power... the list goes on. The key for this step is to focus only on areas that are already of interest rather than ones that are not currently impacting or critical to your business. The idea is not to create an exhaustive list but a focused one.

Step 9: Use the TBD+ Compass to evaluate new areas of technological interest

Now that you have the areas of potential future interest for the business you need to evaluate them in order to create a prioritized action list. You may have already noticed (as most people do) that the list you have created is long and looks unwieldy – do not worry or think about lobbing anything off because the initial list is meant to be long. The process gets quicker after the initial Investment Grid has been filled out.

THE TBD+ COMPASS
The TBD+ Compass is a measurement diagram that will, once completed, show you how the technology you are evaluating fits into your business's future. The diagram is loosely based on a network spidergram and when filled out will create a variety of shapes (iceberg, reverse iceberg, fox, pacman among others) which denote different potentials for the business based on potential for competitive advantage, first mover potential and a variety of other factors.

Here's what the blank TBD+ Compass looks like:

FIGURE 5.2 Blank TBD+ Compass

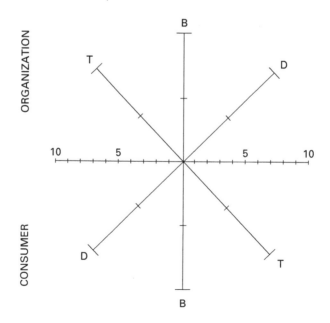

Once you have this drawn out you need to fill it in. Here's what a filled in TBD+ Compass looks like:

FIGURE 5.3 Completed TBD+ Compass

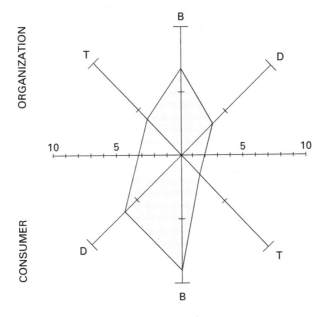

FILLING OUT THE TBD+ COMPASS

This process takes a few goes to get down to a quick procedure but once you and the group have done it a few times you really can see the power that TBD+ possesses to move a business's direction.

How to score   Each section of the TBD+ Compass is for a specific member of the selected steering group.

The T-group looks at the technology itself and the properties and uses and is focused on 'Can they do what you are asking?' based on your business objectives:

- TC – Senior Advertising/Marketing/Consumer-level executive.
- TO – Senior IT Manager/Director-level executive.

The B-group looks at the behaviour the technology elicits, the why behind the technology and how people react to it. Focused on 'Will they do what you are asking?' based on your business objectives:

- BC – Senior Customer Service/Customer Experience-level executive.
- BO – Senior Sales Manager/Director-level executive.

The D-group looks at the data behind the technology, the how many, the when and the how often. Focused on 'Will enough of them do what you are asking?' based on your business objectives:

- DC – Senior Data Officer /Consultant/External third-party adviser.
- DO – Senior Analytics/Finance/Data Officer-level executive or agency.

Every member of the group has 10 points to award or hold back based on research that they do on a certain subject. You can of course feel free to modify the descriptions of the 10-point scale based on your business situation but it must remain consistent to allow fair comparison.

Important note: While money is important, the right answer for your business is paramount when it comes to TBD+. This means that while technologies can be expensive, this factor should alone not exclude them from consideration. The technology that is excluded on this basis could be the very technology that differentiates your business from your competitors' businesses. To this end, remind people about this all the way through the process. It isn't making light of the issue but it is freeing people to think big and see what is possible, not what is immediately feasible.

The 10-point scale is determined by each member (or mini team) answering their designated question:

- TC – On a scale of 0–10, with zero being 'technology doesn't exist' to 10 being 'everyone uses it', how much of the market has the technology?
  - How many people have the hardware, software or application in their hands (or access to it) right now?
- TO – On a scale of 0–10, with zero being 'the business has nothing it needs to roll out the technology' to 10 being 'it has everything it needs to roll it out', how able are you to deploy the technology?
  - What are the current development capabilities of the business?
  - What are the current human resources that will be required in the business that you don't already have?

- BC – On a scale of 0–10, with zero being 'behaviour is not currently happening' to 10 being 'it is already an everyday behaviour', how many people are doing what you need them to do?

  - Beyond current behaviour, are there any complementary or similar behaviours that can be easily applied or modified to make the desired behaviour happen?

- BO – On a scale of 0–10, with zero being 'not keen at all' to 10 being 'incredibly keen', how keen are you to encourage (or cater to) the behaviour?

  - Is it commercially advantageous to drive this behaviour or financial death if you don't cater to it?

- DC – On a scale of 0–10, with zero being 'no interest' to 10 being 'everyone showing interest', how many people have expressed interest in doing what you need them to do?

  - Are people interested in the technology or the problem it solves?

- DO – On a scale of 0–10, with zero being 'not at all' to 10 being 'total disruption', how disruptive will this technology be?

  - How soon is the technology likely to be adopted by the mass market?

  - How pervasive is the technology likely to be or predicted to become?

  - Are any competitors currently in the space or entering the market?

This is a key part of the process, so make sure everyone understands their question and can justify their scores. Once a person (or team) has finalized the score for their part of the compass, have them send it to you.

Step 10: Calculate final TBD+ score for new areas and plot these on the Investment Matrix

Once you have all six scores you need to add them up into a single score.

TC + TO + BC + BO + DC + DO = TBD+ score

At this stage there may be some discussion around final scores when they are seen against each other. This is only natural and is a big part of the process. The key here is to have an open and honest discussion as to why the numbers in question are what they are. The best advice from years of doing this is to spend some time on this area and make sure concerns are explored and ironed out before moving on. The score is a guide and not the final, final score until there is agreement – without agreement you risk losing enthusiasm and support for the future of the investment (and the programme).

CASE STUDY

Here is an example of a finalized TBD+ Compass for a supermarket chain that is looking to explore selling 3D-printed food in their stores.

FIGURE 5.4 Completed TBD+ Compass for supermarket chain

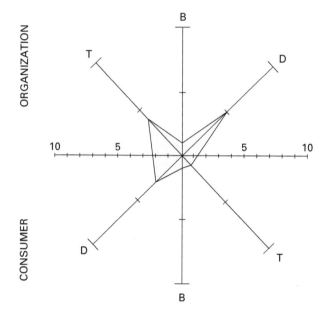

Currently the supermarket is unlikely to invest in this technology because while it exists, it is still early days (TO is midway) and consumer tastes and behaviours for this food need to develop (BC and DC are low scores). In this case, according to the Investment Matrix of the supermarket, a TBD+ score of 15 means no immediate investment but further research is warranted; immediate trial potential is denied but, based on further research findings, funding in the future could be possible. Re-evaluation should begin in six months.

Once this process has been finalized and you have all your final TBD+ scores, plot all the technologies on the Investment Matrix and make sure everyone has a copy in order to maintain cohesion, transparency and keep everyone focused.

Now stand back and look at the Investment Matrix:

- Does anything jump out?
- What are the surprising results?
- Is anything immediately impending?
- Are there any quick wins?
- Is anything set to derail you if you don't do anything?
- What's missing?

Thinking and answering these questions will help you prioritize your first practical next steps. Spend some time evaluating the grid or list in front of you and run different scenarios – are there any combinations? How can you make the most change in the least amount of time? Is that even the right play?

Step 11: Identify key areas of opportunity

Naturally, the higher-ranking numbers will interest people the most but these are often not the easiest or fastest to achieve and often have investment decisions attached that mean lengthy next steps (which is fine). However, don't be fooled into just focusing on the top right quadrant (if you have the technologies in a grid) or the top few lines (if you have listed them in descending order). Prioritizing is key here – look for over-indexing data, rapid growth spurts, purchase-intention data and future potential data points and mark these down as potential interesting points.

Step 12: Determine action plan based on new Investment Matrix

Action plans can take various forms and formats – I prefer to keep it simple and short with clear actions and owners but feel free to use your organization's style.

Whatever style you choose, be sure to include the following points:

- What actions or changes will occur?
- Who will be in charge of carrying out these changes?
- By when will they take place, and for how long?
- What resources (i.e. money, staff) are needed to carry out these changes?
- Who should know what/when?

Step 13: Feed back to the group

Initial feedback should be short, decisive and excited. For best results, we recommend completing TBD+ at least every six months (quarterly is best) in order to a) keep up with new disruptions and emerging technologies but also b) keep you and everyone accountable. The process should be a one-two punch in the first 12 months. Tony Robbins is famed for his weather analogy: it is smart to set your course and course-correct because the weather can change.

TOP TIP

Set calendar reminders for the next TBD+ update early so people are reminded to keep on track and stay focused on the process. Two to five of these are usually sufficient; more than that can cause apathy – you are looking for nudges rather than shoves.

Conclusion

In conclusion, the TBD+ process is a tried and tested way of embracing uncertainty, applying a method to changeable elements and creating a bespoke plan for your organization. Over the next few chapters we will look into some other elements that will help make this process and others a success.

The next chapter discusses an important area – selling back the action plan and how to excite the senior team so they get behind it quickly and without extra work.

06

How to get sign-off

Once you have your TBD+ finalized (or simple TBD if you used that) you will likely have a decision that requires some sort of approval process or sign-off. Disruptive technologies often pose greater and unique issues for investment requirements because of their nature or the uncertainty attached to them. In this chapter, we will explore a variety of issues and exercises that will help and enable you to give your requests and budgets the greatest chance of success.

Why people resist ideas and solutions

Radical and non-radical ideas can be a struggle to get signed off for various reasons – understanding these reasons can increase your success rate and improve future pitching. The most common offender is lack of understanding about the idea itself; either the presentation is poor or the idea is too complex and the company would prefer to defer and give a quick 'no' rather than challenge and exert more energy on it. The second offender is that they cannot see (or are not told) how to apply the idea or the change that is needed. Often this can be overcome with dialogue but depending on the complexity it can often be a tough process.

Despite numerous schemas (mental shortcuts we use to make decisions), people are not programmed to say 'no' to new ideas but we are hardwired to put our world into order. We couldn't survive in everyday life without a system to organize the world into different buckets – mental or social, for example. Gillian Tett, anthropologist and author of *The Silo Effect: The peril of expertise and the promise of breaking down barriers* (2015), believes smart people make dumb decisions for a wide number of reasons but the biggest is often because we fall foul of stereotypes, tribal thinking and risk avoidance. Battling these – and other – ingrained issues can be tough but there are ways to change and alter culture to be more risk-aware instead of risk-averse. Being risk-aware is the ability to take calculated risks and agree to move forward rather than simply stay where you are because you haven't found a safe position to move to. When talking to senior executives, it often helps to discuss having a risk-aware mentality as it demonstrates you are aware of the bigger picture, which puts people at ease. In the following chapters, we will look at creating open businesses and departments – understanding and predicting can often save a lot of disappointment, stress and poor outcomes.

Working in-house and externally in agencies has provided me with a clearer understanding of both sides of the table before, during and after negotiations. These experiences have formed a lot of the way I approach decision makers, negotiate and present information to different parties. Every business is different – despite some key similarities and focuses – but each will have its own way of saying 'yes' or approving projects. Spotting these intricacies and exploiting them will help you maximize your likelihood of success when asking for resources and finance.

Two key areas often stand out: risk and bias. Understanding these areas and how to combat preconceptions greatly increases your chances of success.

Understanding risk and why it isn't a dirty word

Disruptive technologies are by their very nature risky; untested, emerging and sometimes using new areas of technology that require a leap of faith. A healthy risk-aware mindset is key when selling ideas and plans to senior decision makers.

The difference between being risk-aware and risk-averse is a simple one; risk-awareness is an *active* exploration of risk because you believe that higher risk equals higher returns, whereas risk aversion is a choice to avoid risk and actively withdraw from risky opportunities and scenarios. If a company is risk-aware it does not mean that it actively accepts every risky new idea but that it explores them significantly before adoption. Different strategies are right for different businesses and at different times of a business's life. The idea is not to prevent risk – this is impossible – but to minimize it as – and here's the important part to remember – you can't get rid of all risk.

Any business or organization has brushes with risk on a daily basis – the size, structure and policies of an organization help determine the culture surrounding risk. Looking ahead, and to some extent back to recent history, big gains and historic products have been created because of a focus on risk rather than turning a blind eye to it. These companies (Apple, Uber, Meta, Netflix to name but a few) have embedded risk management into policies and procedures throughout their organization – some from the top down and others from the bottom up. Embedding risk management – at any level – implies a company's culture values and considers risk rather than being thrown by it. In essence, it becomes a normal part of the mix and not a rogue element to be feared.

Risk can be embedded in a few ways – this is a long-term sell-in tactic and requires help from different levels:

- minimizing a 'blame' culture by rewarding risks that were taken in the right way;
- aligning individual goals with those of the organization;
- including risk management wording and responsibilities within job descriptions;
- publishing risk stories – like case studies but specifically about risk and how the right decision was made;
- altering new employee on-boarding materials and presentations to include risk wording and examples.

An interesting point from a report produced by risk management firm CEB Global surrounds the future of risk and where financial impact will most likely come from. Historically, risk management is mainly focused on legal issues, compliance with regulations and financial reporting but between 2005 and 2015 most of the big hits to shareholder value actually came from strategic and operating risks like failed internal processes, personnel issues and external events (CEB, 2016).

In other words, the days of risk being a simple thing are fast departing from most businesses, old and young, big and small. Instead, a new era of risk awareness is ushering in a new focus on the strategic issues of risk aversion rather than the easier box checking of other areas. Now is the time to make real changes so you and your organization's future journeys are that much smoother.

Often individuals assume their businesses are more risk-averse than they actually are because of perceived historic decisions, stereotypes and the fact that it is often easier to see why others would not want to do things rather than why they would. Understanding the motivations of people who may push back is important to maximize your chances of success.

Bias is everywhere and nowhere

The world of business is fraught with cognitive biases that hinder great ideas and promote bad ones. From self-interest to making good decisions, it's a dangerous world out there and spotting issues as they arise is a huge part of securing a 'yes' for disruptive technology investment. Below are some of the most common biases that behavioural economists and psychologists have identified; being aware of these biases can improve your pitch and increase your chances of success:

- Confirmation bias – where people give extra weight to evidence that goes along with their existing ideas and beliefs despite evidence that contradicts it.

- Anchoring – people root their decisions on initial feelings and fail to move beyond or explore other decisions.

- Groupthink – people look for consensus over more judicious appraisal of ideas.

- Loss aversion – people fear loss more than value gains of the same amount.

- Vividness bias – people focus less on poorly presented data but pay extra attention to vivid information.

- Sunk-cost fallacy – people focus on costs that have already been put down and fail to take different courses of action despite poor results from the current path.

- Existing commitment bias – when extra resources are poured into a poor decision because of the existing money, time and additional resources already invested.

- Present bias – people value immediate gains over long-term gains.

- Mere-exposure bias – people develop a preference for things because they are familiar with them.

- Status quo bias – people prefer to leave things alone without the pressure to change them.
- Overconfidence bias – people overestimate their own skills and their ability to affect different future outcomes the further out they seem.

Biases are often deep rooted and difficult to shift – remember the goal is not to change the person but guide them through the process towards saying 'yes' to your plan. Each of the biases can be reversed and used to your advantage – remember to *help* them make a decision, don't railroad them into making one:

- Ditch the data tables – make data vivid, bring it to life for people (vividness).
- Call out the difficulty early in order to explore and move past it (status quo).
- Use short-term gains as a means to sell in long-term gains (present).
- Focus on existing project costs to free up sums of money (sunk-cost).
- Play back what will be lost before what will be gained (loss aversion).
- Regularly exposing disruptive and emerging technologies in various ways will increase acceptance and preference for similar ideas (mere-exposure).

This last point is key; exposure reduces fear and also opens up minds to the possible. Show and tell is when by far the biggest 'aha' moments occur when talking about disruptive and emerging technologies to the most technologically challenged individuals and organizations. Before the meeting (priming psychology) you may want to organize a Trend Safari. This exercise is a day out of the office for a select group of decision makers to experience and absorb different new technologies and concepts and then report back to the wider team.

ACTIVITY TREND SAFARI

Time required: 4–8 hours.
What you will need:

- three organizing people;
- five to ten (max) attendees;

- fresh notebooks and pens for attendees;
- an end-of-day space to wrap up learnings and relax/reward.

As the name suggests, a Trend Safari is an activity that you go on and experience rather than just going through the usual show and tell mechanics. The locations will vary (which is why careful time planning is essential) and different activities take place. Some will be hands-on with new technologies, others will be Q&A sessions, and some may be observation sessions where you simply observe your consumer in the wild. The point of a Trend Safari is to create an immersive experience that brings technology and changing consumer behaviours to life. The role of the trend safari can be entirely inspirational as much as it can be used to prime the 'yes' pump. In terms of the group, 10 is usually the biggest. More than this and time gets lost with toilet breaks etc; in addition, many venues will not have the capacity to hold more than 10 because of the stage the business is at. The safari should feel action-packed (with sufficient buffer time between venues) but the idea is to leave people wanting more after each location. Building in talking points or even exercises is a great way of maintaining momentum – this also encourages participants to reflect on insights and actions to take away. Bespoke materials such as booklets with information about the locations and space for notes will also enhance the overall feeling of a carefully tailored experience.

Locations/vendors: Finding the right vendors and venues to visit is key. There are numerous ways to do this depending on where you are in the world and how advanced the technology is you are looking at. StartupBlink (startupblink.com) is a global map of start-ups around the world that you can use to zoom in on interesting start-ups in your area.

When choosing locations for the itinerary, there should be a strong case for each venue and its contribution to the overall goal of the day, whether organizing a safari further away or in another market, or operating on more familiar ground. Once you have identified a number of interesting locations, explore the surrounding area on foot to spot any additional options – disruption and innovation tends to be fertile ground for more disruption and innovation. Don't forget, live webcasts could be another option if the company you would like to see is not in your area – although this should be a last-resort option.

Other places to look for inspiration on who to visit include *Wired* magazine, Nesta.org, MIT, Techcrunch ('Disrupt' conference is particularly interesting), XPRIZE, TED, Investopedia and good old Google searches.

TOP TIP: To complement your choice of locations, consider including briefings with relevant speakers on the itinerary, for example during lunch or a coffee break, so the meeting location could be linked to the overall purpose of the day (e.g. a restaurant that boasts interesting payment technologies or has robotic waiters) as a compelling backdrop for the guest expert.

Budget: Depending on the venues, transport, lunch and speakers you should aim to spend between $2,000 and $10,000 on the process. Some clients have used a minibus to get from venue to venue whereas others have opted to take Ubers or simply walk if this is possible.

Note: Trend Safaris are not 'agency days'. Such days are often slanted in favour of specific goals and are not necessarily going to show you disruptive technologies. The beauty of the Trend Safari is the ability to break out of the usual echo chamber you are in and hear fresh and exciting perspectives in addition to seeing the latest and forthcoming technologies.

The Foresight Factory (a global consumer trends agency) runs Trend Safaris regularly for FMCG, travel, and retail brands. Co-owner and CEO, Meabh Quoirin, has a series of tips for creating the best Trend Safari possible – whatever your budget:

- **Start and end with the business**. 'Link the safari's itinerary to existing projects, issues or trends in focus to drive further action. Linking also provides more ongoing value by sparking new ideas and triggering insights to follow up on versus the more inspirational outcomes.'

- **Look beyond your nose**. 'We select experiences that are likely to be fringe/emerging now that will develop into more mainstream opportunities. We want people to feel the future before it happens. We want them to get to the future before the competition.'

- **Communicate the value early to senior members**. 'Often time-pressed, senior staff may wonder if the safari will introduce them to anything new or useful. Communicate the value of a safari and frame the day in the context of commercial goals and initiatives along with additional benefits like practical experience ("you might know about these innovations already, but you have not actually tried them yet") and the opportunity to bring together different departments within the company which may usually operate in silos.'

- **Plan time to plan time**. 'Safaris are difficult to wrangle – even with existing relationships. Ensuring availability and meeting up with key contacts at the different venues can easily take several weeks.'

- **Confirm and re-confirm**. 'In the days just before the safari, when you are finalizing last-minute practicalities and materials, double-check the various locations. Is the agreed timing still suitable for a visit? Is the contact at the venue still available and do they have any more questions ahead of the day? The organizing team should also run through the day's "screenplay" together and identify any possible pain points.'

- **Plan for the future**. 'Keeping an ongoing database of potentially interesting locations and start-ups as you identify them (or tapping into the database of a company working in this space) will also create efficiencies for organizing future safaris. Often consumers themselves are an invaluable source for locations. The Future Foundation has a global network of Trendspotters which keeps us primed regarding newly launched venues and activities in over 70 countries. The people in our network have unrivalled knowledge of their locales – the coolest retailing, the hottest bars, the newest leisure venues.'

TOP TIP: Think about whether there is value in a 'reveal', i.e. not showing the full itinerary to the whole group until the day itself, using 'teasers' to whet appetites instead.

Outsiders are often seen or needed to create movement

Sometimes you need external help, either with implementation, ideation or identification of possible solutions. The latter can often be the best route to a 'yes' as external experts are often given more boardroom clout based on their presumed – and actual – expertise. While direct expertise and experience are always useful, a new trend is emerging called 'analogous-field thinking', which helps companies move forward with new technologies using thinking and experiences from other fields. This idea is explored more in the next chapter as it helps to open up organizations and thinking patterns.

An alternative method to help move things along quickly is to remove or reduce the friction in the process of getting the 'yes'. Most organizations I have worked with have had at least three extra steps before sign-off that could be removed or reduced – this could be the difference between effectively stunting creative ideas and gaining investment to explore new opportunities. Streamlining your 'yes' process before you ask might be the smarter move before an external consultant is brought on.

Selling power... painfully

We are all increasingly selling to different parties whether we are negotiating with a toddler, a barista or a boardroom executive – each requires different selling techniques and focuses. 'People buy emotionally but make decisions intellectually', according to Lisette Howlett, Senior Sales Trainer for Sandler (a professional selling programme). 'In order to really effect change or get a person to make a decision, that person must be in some sort of emotional "pain". In other words, he or she must understand what this decision will mean for them.' By doing this, you reframe the decision from what is best to something the person needs to protect to something they strive to achieve – a powerful motivator.

This technique is known at Sandler as 'The Pain Funnel' and describes a series of more and more focused questions that take the decision maker into their world rather than the role they are being paid for. To be clear, this is never physical pain – rather emotional triggers that help you understand the person's motivations for making decisions that can affect you. In other words, the decision is not 'should the company invest in 3D printing?' but instead, 'what will investing in 3D printing mean for my desire to get out of work an hour earlier each day so I can put my children to bed?'

Howlett suggests five tips when selling in large projects regarding disruptive technologies.

1 Sell today, educate tomorrow

Creating a slew of materials upfront might not address what is needed by the decision maker. Success usually lies in not creating anything (or too much) upfront and instead interrogating the commitment to the project. Focus on qualifying the decision maker's commitment by asking open questions along the lines of:

- The last project you tackled like this one – how was it?
- We've previously discussed [ISSUE]; is it still an issue, have we given up trying to fix it?
- What would you need to see in order to support this?
- Would else would you need to see to be satisfied?
- If you had a magic wand, what would you do?
- Similar projects have cost between [AMOUNT 1] and [AMOUNT 2]; what end would you be comfortable spending?

2 Be a dummy

This exercise is great to get others talking. Instead of going in as the expert and talking at the decision maker, hold back sharing everything you know and ask smart questions so you can gauge the person's interest and commitment. Amateurs blather on about things they think they know about. Professionals know what they know and how to find out more. While being a dummy may sound counterintuitive, it is often a useful selling technique to help people feel empowered and do most of the talking.

Use phrases like:

- 'Did you mean...?'
- 'Let me see if I have this straight...'
- 'Tell me more about...'
- 'I don't suppose...'
- 'I'm confused...'

3 Prevent buyer's remorse before they buy

Big projects can be pulled at the last minute when people reflect on them; it is important to prevent this in a few ways. Double-checking the decision maker is happy with everything is a good start but going beyond this and talking about their greatest fears at this stage often surfaces issues that can be quickly mitigated against. A great way of testing this is to use an imaginary thermometer. 'On a scale of 0–10, with zero being "incredibly unhappy" and 10 being "blissfully happy", where are you with our plan?' Now you likely have some wiggle room to discuss, call them on it and mitigate against it.

4 Monkey's paw

Based on a nautical term, the 'monkey's paw' is a small ball of rope that is attached to a larger line enabling the larger line to be pulled in – without it, pulling in the larger line would be dangerous and near impossible. This technique is useful when decision makers are totally capable of saying 'yes' but are not ready to commit fully. In this scenario, an all-or-nothing approach won't work so you need to complete two separate 'sales' (or decisions). The first sale will be a pilot programme or initial step and then the second sale is the rest of the programme based on the success of the initial one. Completing the initial project will give the decision maker greater confidence and transform the next ask you may have.

5 *Blow up the bomb early*

Howlett's biggest tip is owning the elephant in the room or, as she describes it, 'blowing up the bomb early'. Disruptive technologies can make people uncomfortable and come with a lot of caveats, unknowns and costs. By being upfront and putting forward your concerns (as well as the opportunities) you can move beyond the fears and concerns and really explore potential and likely outcomes.

'That's expensive'

As previously mentioned, potential investment should not have limitations placed on it and until now, both simple TBD and TBD+ have downplayed the financial aspects of any change looking to be made in the exploration and refining phases.

Money often has a polarizing (and negative) effect on research and outcomes because of perceived biases, stereotypes and personal history.

Another reason for this initial blinkering is based on experience and history of implementing change within organizations of varying size; the best solution is still the best solution even if you cannot afford to implement it. Beyond this, of course, money for good ideas usually has a way of being found. Now that you have put your plan together it is important to think about the various costs involved – financial, human, social and so on.

Disruptive technologies are often perceived as expensive. Every company and person is different and the figures you have in your head may be drastically different to the ones the decision maker has or has access to. Assume nothing when you go in and instead frame the position or positions you want to propose.

The good thing when talking about money is that you have lots of options, from blowing up the issue early (see above) – 'It's a given that it will be a decent amount of money…' – to bracketing – 'What would you feel comfortable investing in this area?' – or reframing the issue entirely – 'Money is unlikely to be the real hurdle here, what I'd like to talk about is time…'.

TOP TIP

Change the thought patterns with different wording. Instead of using these words and phrases, try using some alternatives and observe the different types of follow-up questions and responses you get.

TABLE 6.1 What to say, what not to say

Use	Don't use	Example
Money, invest	Budget, finance	'Is now a good time to discuss the money for the project?'
Decent	Significant, large, big	'It's a decent amount as you might expect.'
Comfortable	Willing, able	'How much money are you comfortable investing in this project?'
Should	Could	'We might do this in lots of ways but I wanted to get your take on what we should do.'
Under	How much, the cost	'And you'd be hoping to bring this in under how much?'

The bottom line of this part of the process is a simple and sobering one: you can't lose what you don't have. Be bold, keep your nerve, don't baffle with jargon or too many numbers at once and start to bring them on the journey by seeing the potential and the risks in a cool, calm and collected way. You're in this together after all. Go for gold.

I have three rules when pitching something or if someone is pitching something to me.

1 Help me know your product (even if you don't fully know it yet)

This tip originates from the start-up circuit and is often bandied about at hackathons, start-up conferences and investment competitions. Great pitches have some key characteristics which other experts and books focus on but without a doubt the most successful pitches I have been involved in had one thing in common: the presenter made the target understand the product – how it works, why it is needed and what drives it.

TOP TIP

Present to your least technical or tech-savvy friend and ask if they understood what was being sold, the reason and what was being asked for. If the answers to these aren't what you need, fix accordingly.

2 Less is more (but make sure you have the more to hand)

Thinking back to the times I have been a part of or have had to sit through decks that are ridiculously long (my record is 203 slides) makes me sad. Not just because it is time I will never get back but also because it shouldn't be this way. Most things are not that complex – so why make them complex? People believe longer and bigger means money will magically appear in their pockets – this is not the case. Often, complexity over brevity simply confuses the information recipient or just creates an inability to make a clear decision. Neither scenario is desirable, so think about what you need them to know versus what you want them to know. This is usually about a 1:10 ratio. Ask yourself, if I was on the receiving end of this presentation would I thank them for this information at this stage?

TOP TIP

Review your presentation and ask yourself what every slide is adding and why it is there. If a slide isn't adding something vital, remove it, keeping the information in an appendix or notes section. The information isn't bad or worthless but right now it is harming your chances of success because it is distracting, detracting or damaging (the three Ds) to your message or desired outcome.

---

3 Pre-mortem a change

This is a technique I use when I know I have a client who prefers the big picture or likes to see the goal before the procedural stuff. I fall into this category and many top decision makers and C-suite executives have this characteristic too. The idea of a pre-mortem is the reverse of a post-mortem where you try to understand what happened. Instead, with the pre-mortem, you focus on a possible failure and work backwards to identify potential issues ahead.

Often we get so blindsided by what we are trying to achieve that we forget the simpler, subtle things that help us sell and get what we want. From my days studying psychology (and keeping up with behavioural economics research), some recent findings have application when selling ideas and projects.

TOP TIP

Start with a 5–10-year scenario statement along the lines of the following: 'It is 2026 and our distribution costs have skyrocketed by 150 per cent – why is this percentage so high?' Doing this enables you to pre-empt issues and possible questions the decision maker(s) may have. In addition, this technique also enables you to play to the ego of the decision maker and appear to forecast the future while offering up suggestions for possible routes to take.

BLOOD SUGAR HAS A MAJOR EFFECT ON GETTING A 'YES'

Studies have shown that judges gave lighter sentences after they had eaten, and the same goes for most people making a decision. More positive outcomes or responses result from a full, not gurgling, stomach. Book in the meeting after lunch or a mid-morning (or afternoon) snack so sugar and energy levels are high.

SHUT YOUR MOUTH FOR A SMARTER RESPONSE

TED speaker and author of *Gravitas*, Caroline Goyder, is a master of helping people to become more confident and successful. In her now infamous TED talk (Goyder, 2014), she discusses the power of 'breathing low and slow' as the key to speaking with confidence and clarity. It is the 'in breath' that we take when our mouth is closed that enables us to think before we speak and also focus on what we will say next.

Conclusion

This chapter is a crash course in selling difficult concepts, new ideas and unknown entities. The list of techniques and resources available to you beyond this is vast and I urge you, if you aren't 100 per cent confident in selling, to explore and develop this skill further before diving into the abyss with large decisions that could totally change your company's future.

In the next chapter, you will find out about creating an open business beyond the TBD process.

07

Open business and innovation

In this chapter, you will find out about creating an open business beyond the TBD process. By fostering and encouraging an open business and environment, future investments and explorations into disruptive technology won't just be easy, they will be faster too.

This chapter will show you why open businesses are the future and how to:

- foster an open business/department;
- create a culture of original/open thinkers;
- discover the benefits of open and original thinking.

Disruptive technologies require a different mindset from that which many businesses today display or hold at the centre of their belief structures. Beyond simply being a badging exercise, being more open and transparent is not just about looking different – it is imperative for the external and internal success of a business. By being an open business, you start to see new ideas, new ways of completing different elements, potential issues and opportunities to leap forward. The pandemic and the following period offer big opportunities for businesses to rapidly change and adopt new ideas and ways of working. Working from home became the norm for many and because of this many businesses now operate on a reduced in-office schedule. The same could be done with management and teams.

'Open business' is a much-bandied-about term, conjuring up images of open-plan offices, glass walls, graffiti, endless scribble on whiteboards and the obligatory casual workers slumped on beanbags with faces lit up by their laptops and tablets. This is a reality for some but the goal for many. This scenario is the ideal and while it isn't right or wrong, it has become what a lot of people strive for or think will solve all their problems. Experience and history tell me that this 'ideal' is far from what most businesses actually need in order to change, survive or thrive when disruption is occurring. Instead of amazing offices and flexible working, many executives I talk to actually long for a more fluid communication and management style. In other words, the exterior that your business creates needs to mimic or take its lead from the interior you have, in order to maximize your people resources.

What is open business?

At its heart, 'open business' means having transparency and accountability at the core of your business – a dedication to every member of the business and openness to new ideas. Building on the idea of open source, Alexander Stigsen invented the open company in March 2009 with his company, E Text Editor. This was later followed by the much-cited Gittip (now Gratipay), where he outlined his three main aims in a plain blog post. The aims were:

- share as much as possible;
- charge as little as possible;
- don't compensate employees directly.

These sound harsher than open business actually is – the employees have access to and get other things that are non-monetary in value. The idea is to benefit society as a whole, not just a small collection of its members.

The concept is terrifying for a lot of people because it is not the way they were taught or ever thought they could work... and yet it is working for companies across the world, and not just start-ups. However, when you delve into open business, most businesses realize they are on the path, already demonstrating a lot of open behaviours.

There are several key areas that open business addresses:

- *Principles*. A focus and agreement to exchange insights, knowledge and findings using open source, open standards and other open technology methods as much as possible.

- *Sharing knowledge*. This is fundamental to open business – a key focus on sharing knowledge and learnings at all levels and locations. This element was a driving force behind the way TBD+ was designed to include the majority of the business leads. Simon Sinek, bestselling author, agrees, and when I asked him what one thing he recommends businesses do to avoid disruption, his unhesitating response regarded this area:

> The businesses that will outdo their competitors in the next few years are the ones that develop comprehensive leadership training. It is a big gaping hole in

companies today. It will improve the overall quality of the kind of skills in the people for whom we work and we will all feel that the company cares about our development. (Sinek, 2017)

- *Finance.* All members should be able to see the accounting details and compensation of others. This is often the most controversial element of the movement.
- *Participation.* A key part of open business is that everyone participates.
- *Open individual.* Each member of the business is encouraged to explore their personal development – technical or spiritual.
- *Community.* Non-business-related activities (religious events, family commitments) are seen as important to the success of the business because the individual is satisfied.
- *Access.* Each member should have access to the contact details of all other members – whatever the location (once approved by the member).

Reading the list may sound anti-capitalist and this argument has been lobbed at the open business movement many times (and with good reason). Lack of transparency and hiding behind the need for corporate secrecy are major causes of upset with employees and external interested parties, yet both have been a part of business for decades. The open movement is not law, doesn't ignore the intellectual property of a business and really aims to improve trust – internally and externally. If you approach it – as many big businesses do – as something that should be feared or avoided, you will get poor results. Instead, I urge you to embrace it as an opportunity to improve and create more value from assets you already have; you will see very different results. The speed at which you embrace openness is going to be different for every business – but as more and more workers list 'openness' and 'trust-worthiness' as key elements they look for in a business, you might be disadvantaging yourself in both the short and long term by not being more open. For more information and resources about open business, check out the Open Business Council (www.openbusinesscouncil.org).

Open business does not mean (or require) Holacracy

Many people mistake open business for Holacracy – a business management style that flattens (note: not totally flattens) a company structure and morphs people into roles and groups instead.

Emerging partly due to the public relations efforts of companies who are adopting it (Medium, Zappos – owned by Amazon – and government departments in Australia, New Zealand and others), Holacracy is a new style of organizing business, but it is a management style rather than a larger movement like open business. With strong pros and cons, Holacracy is an interesting idea that will work well for some businesses but could also completely destroy others. Whereas open business is about knowledge and understanding, Holacracy utilizes a clear set of rules and processes for how teams break up the work with clear roles and responsibilities, and without micromanagement. In practice this looks a lot like most of the offices you have ever been in – 'cool' or not. Holacracy is simply small groups in the same space that may or may not all interact with each other in order to complete specific tasks and streamline the process of doing whatever they do, whether that is offer a service or create a good. If you are looking for more examples of Holacracy look at companies like Zappos, Medium, Decathlon, Starwood Hotels and Resorts and Kingfisher among others. Some of these companies have fully adopted Holacracy while others have limited it to specific departments as you do not have to completely adopt the system all at once for it to be effective.

Sceptics claim Holacracy is difficult to really ever implement with established companies but while it is unlikely to see mass adoption, there is an important idea at its heart. Most businesses are not capable of such a drastic change (although it may be exactly what's needed) and many management styles or ideas never progress after the initial period of excitement (Google's famous 20 per cent time, for example). Instead, a simplifying of business structure and a more thorough job description of roles and responsibilities may be what's required to gain the benefits many say Holacracy holds. This is why open business is often an easier sell to companies.

So how open does an open business have to be?

Being truly open might be an unrealistic option for many companies due to the sizeable financial and productivity ripples it could cause in the short term. Instead, history tells me a more fluid approach is recommended for most businesses. Open business does not necessarily mean everyone has access to all information about salaries etc, although this is a goal of the Open Company Initiative (www.opencompany.org), a group of companies who are genuinely committed to open business and creating trust.

The OCI website states:

We envision a world in which companies are generally trustworthy. We believe that the winning companies of the future will be those that proactively maximize trust, going above and beyond the letter of the law to cultivate relationships of trust with customers, stakeholders, and society as a whole.

Companies that have embraced open practices have the freedom to make decisions, take chances and change strategies without fear of reprisal but with clear direction and organization. This is a nice sentiment but there's a lot of legacy baggage for most companies when it comes to financial rewards in particular. Therefore, a phased approach is recommended if the 'ripping the band-aid off' approach is not possible for your company. This chapter deals with the realities of open business and gives you a practical guide and set of points to think about when making your business more accountable and transparent. Change starts with you.

Why is being open such a big deal?

As we'll discuss in later chapters, younger members of the workforce and future workforces value openness and transparency so, in order to save money on recruitment and to recruit the best talent, it makes good business sense to take openness seriously. Beyond talent, though, what are the other benefits for a business?

- *Problems are found and fixed.* Instead of being kept secret, hidden for fear of blame or finger pointing, ideas and issues are sought out rather than shied away from.

- *The sum is greater than its parts.* Working openly fosters and enables greater products to be made because people are working on something larger than they could have created on their own.

- *Better customer relationships.* Building trust is a key reason more businesses 'go open'. By providing customers with more access to data and documents there is less ambiguity for people to feel uncertain about. This has enabled companies to create better-quality, more loyal and more honest early tester communities (IBM, Zappos, Microsoft).

- *Lower risk.* Being open means employees have the right information they need at all times, thus lowering risks when it comes to the product and personnel.

- *Co-creation.* The pool of ideas will always be bigger outside a business than within it because of the numbers involved. Harnessing these ideas in a smart way maximizes resources internally but also gets the best from external sources.

- *Lower costs and faster orders.* Lower costs for development and for introducing products to new or existing markets have been shown thanks to the centralized approach and the ability to make multiple decisions easily. Additionally, experts and users can be accessed quickly thanks to the organization structure, communication tools and community approach; orders can be processed faster because the business has centralized intelligence on each order.

CASE STUDY
Connect and Develop by Procter & Gamble

What

An open innovation project that licenses and acquires products from other companies to bring them to market as P&G brands. Early success was seen with successful collaborations with the skin cream Olay, toothbrush brands and cleaning products under the Swiffer brand.

How

A bespoke submission site enables anyone with access to send P&G ideas and submit product improvements. P&G then have a dedicated team who review and evaluate every idea (and keep the owner up to date). Additionally, P&G use an extensive network of scouts to identify and evaluate external ideas and technologies that could be applied or added to P&G brands. P&G have an annual awards show for their scouts and have also partnered with several universities to get first looks, propose bespoke problems and ideate from new and emerging designers.

Results

More than 2,000 successful agreements with designers and innovators in the first 10 years. By adopting a more open and welcoming approach to innovation and disruptive technologies, P&G have created a positive and easy system for inventors and people with good ideas to be heard and acted upon swiftly. Beyond additional revenue from new brands and sustained or increased market growth for established brands, P&G have created an easy-to-maintain system (with more than 7,000 virtual and 'extended' partners) that is impacting their bottom line and is a powerful resource for the future.

For more information check out the case studies on P&G's 'Open Innovation' site: pgconnectdevelop.com.

A halfway house is possible... but risky

P&G aren't an open business in the truest sense of the term but neither are they alone in their open approach when it comes to disruptive and emerging technology: IBM, Xerox, Ford, Dominos, Lego, Samsung, Dell, Starbucks, Cisco, SAP, Microsoft, Google, HP, General Mills, John Lewis, Barclays, Fujifilm, Unilever, Nestlé, Marks & Spencer, LinkedIn, Ericsson and hundreds of other companies are all using outside forces to help them innovate and be less impacted by disruptive new technologies. Agencies are also launching 'labs' and hubs to help clients navigate new arenas, although these are often vanity projects. I have explained in the *Guardian* why 'F-labs' or fake labs don't work and offered some helpful advice for brands looking to go down this route (theguardian.com/media-network/2015/jun/18/agencies-innovation-lead-labs-incubators).

The key lies in the application back at the base – most initiatives fail because what happens in the labs or in the department can't be brought back to base and made to work. This is why fully open business, while being the best policy, isn't always applicable and is especially hard when you have a big business.

So if open is hard and halfway isn't great – where do I start?

When I go into a new client there are a few indicators I look for first – not just to gauge the lie of the land but also to inform how I approach different elements of the brief and negotiation. One of the most powerful elements I look at (or ask about) is how freely employees can actually speak about the business and how much they actually do this. The majority of the businesses I work with have 'open door' policies but in reality the doors are often very much closed.

The issue is a simple one; if employees don't feel able to speak to you freely about both the big and the little items, you are missing the ideas that could really push your business forward. Sure, you're likely to get minutiae but you could also get a macro nugget that changes the way your business progresses.

The problem is that employees keep quiet instead of speaking up and challenging people – this holds everyone back and I urge you to beat it from your business with every fibre of your being. Studies have shown that when employees feel empowered to voice their concerns and speak freely, not only does employee retention rise but employers see a dramatic shift in productivity.

It's not an overnight task to make your business talk more freely but it is an important one. Here are some strategic ideas that work in the short and long term, and will help you transition from a closed to an open office or department.

1 Ditch the suggestion box

Anonymity can be used in a lot of situations to get amazing feedback but it has one significant drawback – it implies, almost confirms, that employees are not able to speak freely in the first place. Suggestion boxes can be used but the best results come from when employees are able to speak openly and see the results of their efforts. Beyond this, there are a few reasons why the suggestion box is doing you more harm than good. Some problems need to have details – perhaps a manager is a bad manager or there have been more serious issues. Not knowing this person's name is counterproductive to the process and, because you can't identify the individual, you have to spend money on training everyone. Beyond this, the need to then find individuals and assign comments to them is equally counterproductive.

2 Don't make feedback a 'thing'

Sometimes feedback and little 'chats' can seem forced, self-serving and hard to handle for employees and managers alike. Levels exist and often business colleagues are not friends – there's always history between people, even if that history is neutral. If you treat feedback as a chore and only go through the motions you are disadvantaging yourself, the employee and the company. Instead, try making feedback more 'always on' and flexible with smaller, more casual chats that are regular enough and always face to face. Employing this method builds real trust between people and makes the exchange less stressful and overbearing through building a real rapport over time.

3 Don't walk the floor, know it

Open offices are often just big spaces with offices at the edges. Some newer buildings are really challenging this but the majority still have the basic layout and structure. Remember that sitting in offices and discussing topics – even sensitive topics – raises barriers and implies hierarchy. This is a reality for most businesses but the perception – or cues – can be softened by simply relocating and acting differently.

The psychologist J Richard Hackman called these cues 'ambient stimuli' and they can be locational and physical (Hackman, 1973). A prime example is the cost of furniture; the higher up the food chain you go in an organization, often the furniture climbs in price. If you want people to know they are on the same level or that such issues aren't important, use the same furniture as they do. Beyond this, think about your posture – standing too close or sitting behind a desk with your arms behind your head may be comfortable but make an effort to go to the person you are talking to and match their body language. Often perceived aggression and dominance can be from simple body cues that people unknowingly purvey – be aware and alter elements accordingly for the best results.

4 Brainstorm questions first, not ideas

The average company is often poor at completing high-level brainstorms for several reasons: poor process, the wrong participants, requesting 'blue-sky' thinking and bad timing to name but a few. MIT and Microsoft understood that coming up with big leaps and breakthroughs is rare so a new system was needed. Both organizations opted for a more iterative process where assumptions are challenged before participants get to add to ideas and try to push them forward. This process enables your employees to challenge their assumptions and create a deeper understanding of the issue(s) involved before they brainstorm solutions.

5 Be clear on the results or next steps

The key is to be specific with what you are asking. If it is unclear what you need or want from the other person you risk discarding the interaction, sending the message that the individual's ideas are useless and you wasted their time. Instead, think about the question you need answered and how you can work it into a conversation or – if circumstances require it – how you can get a quick answer without seeming rushed.

The key is to focus on what is required and important because without focus you could do more harm than good if ideas, comments and suggestions are ignored. The best way to do this is a public document, promise or procedure that all employees are made aware of (ideally in a prominent place). One of the companies I worked with actually erected a stone in the lobby of the building that had the process etched into it. After the initial period of a few months, the stone was moved around the building to encourage different departments to adopt the mentality too.

TOP TIP

Being clear on outcomes and processes is absolutely vital at the beginning of new initiatives and big ideas, otherwise employees simply forget to bother to voice concerns, issues and ideas in the future. When it comes to implementing emerging and disruptive technology, a good way of avoiding this fatigue is to create a steering board including members of the business from all levels (and possibly outside sources – see more later). This method has worked well for several clients who wanted to create cultures that were not only open but also effective.

For effective change, you need to create some disharmony

Think back to the change equation: change = dissatisfaction with the present × clear perception of the future × first practical steps. A key part of this is being angry at the present. The issue is that most companies actually often just try to make employees conform. Clients tell me that employees aren't innovative enough and never come up with ideas but actually, when I dig into the culture and business practices, it becomes clear they are asking for the exact behaviour they have never enabled. The company instead opts for enabling a few people or holding limited exercises with the hope that the outcomes will filter down or work without much help or further finessing. History and the speed of change and organization issues we see in the media seem to confirm that this approach doesn't work.

Instead, I recommend that clients open up this approach and rather than focusing on the few 'superstars' or participants recommended by department heads, get a broad cross-section of the business or ideally 'free' them all after explaining that if the business doesn't do this, it can only expect to produce similar ideas and creatively stagnate. Giving all employees the opportunity to generate ideas and solutions will identify issues, new skills and skills gaps to plug. The key is having a good vetting process (also see the clear next steps point above).

The judging process needs to be based on the following criteria for maximum effect and efficiency:

- Transparent – people must be aware of the process being applied.

- Status – the process should let applicants know what stage they are at.

- Honest – not all ideas are good, timely or will work – the key is to have honest criteria and feedback mechanics to deliver both good and bad news.

- Criteria-based – not all ideas should be judged on the same criteria, but each must have the right criteria applied to it so evaluation is fair.

- Celebratory – the process should be public to reinforce additional idea generation in both the short and long term.

ACTIVITY
Kill the company

A great way of creating disharmony is to unshackle employees and ask them to do the unthinkable... break the business. In the following exercise adapted from an e-book by Lisa Bodell (2012) from Futurethink (an innovation training company), employees are given a very specific task that enables them to be bold, think outside of the box, expand the regular brainstorming problem/solution mentality and help the company see new areas of opportunity, threat and prosperity. The reason why this exercise works so well is not because people are negative or hate the company but because it forces them to really open up and look at what's not working. Strategy rarely works this way in any busy business. Instead, planning usually involves coming up with new ways of doing something, a new element or more of a certain element. This exercise flips this thinking on its head and helps people identify what isn't working now to clear space for new concepts or ideas that will.

Time required: 2.5 hours.
What: Brainstorm (in teams).
What you will need:

- a decent-sized room;

- five to ten (max) attendees;

- one to two facilitators;

- wall space;

- Post-it notes or whiteboard markers.

Step 1: Set up the task (10 minutes). In teams of three to four, set up the room with tables or access to the space that you have available. Then describe the brainstorming process to the group and discuss what will happen with the ideas that are generated. Thank the participants for their time and urge them to have fun, take the task seriously and be brutal.

Step 2: Ask the Killer Question (5 minutes). The main task surrounds a statement that groups must then discuss and come up with answers to. The idea of the 45-minute brainstorm is to help people think about new and different ways in which the business could be killed. This is when you can reiterate that you want them to be brutal. Being 'brutal' is a key word here. Mediocre phrasing within this exercise won't produce great results. Make sure you tell participants the gloves are off.

The killer question: 'Imagine you are [insert your number one competitor here], write down as many ways that you can think of that could end our business. There are no wrong answers.'

Step 3: Brainstorm (30–40 minutes). Write the statement somewhere prominent and then let the teams start work on identifying everything that is wrong with the company that competitors could, should and might exploit. Once the killer question has been read aloud – and it can be tweaked (although shorter is better) – it is important not to give examples or frame solutions in any way so you will receive the most honest and widest-reaching ideas.

TOP TIP: A note on politeness and tact – it has no place in this exercise! Instead, make sure participants know that you expect to have some harsh statements, ideas and feedback. Give them permission to 'make the company bleed'. This is both cathartic and also freeing for participants – it also enables deeper ideas, rather than surface issues, to be identified.

The room should be noisy and ideas should flow freely – this exercise is less of a competition and more of a chance to demonstrate ideas and show off knowledge. Give time updates as the time goes on and in the last five minutes recommend the teams review their ideas and make sure everything is captured and can be explained.

Additional idea/option: You can organize people by job, department or function if you believe this will give better results or you have a specific business reason for doing so.

Note: Avoid framing. Resist referring to the ideas as 'big' or rewarding the best idea – all the ideas are valid and no idea should be 'best'. Instead, recognize individuals and draw out major themes; this demonstrates group cohesion at the end of the task. Naturally, some may need an example so think of one before the exercise begins in case it is needed. You may be surprised just how many ideas the group can come up with.

Step 4: Collate the ideas (20–30 minutes). Once the brainstorming time ends, teams then have to discuss the ideas with the rest of the group. Each idea is then put on the wall and grouped together if there is overlap or if ideas correlate (i.e. same department, same issue). There should be multiple areas of overlap.

Step 5: Review the ideas (20–30 minutes). First, mentally review the clusters yourself – are there any themes? What does this tell you or indicate about the business and the culture right now? It is now time to think about possible solutions so, as a group, you now determine the priority of the ideas that have been generated. Severity, cost and time to impact should all be taken into account (make sure people are aware of the final criteria you choose – every business will be different). You now have a choice on how to proceed, to identify what idea (or, more likely, ideas) should be focused on. Using Post-it notes or a Google Form that can be created on the fly (this avoids groupthink and people simply voting where the majority have voted already), every participant votes on the top three ideas they believe are most critical for the business to address first. Alternatively, you can be more scientific and follow the method laid out in Bodell's *Kill the Company* (2012) and use a matrix to prioritize the ideas.

Using two criteria – severity of impact to business and likelihood of happening – draw an axis with 'High' on the top and right points of the axis and 'Low' on the left and bottom axis points.

Now place each theme within a quadrant. The most pressing issues should be put in the top right quadrant as these have the biggest impact on the business and are most likely to happen. Think carefully about the placement of each and why they should be placed in each spot – if you find this difficult, use a scale as you did with the Decision Matrix (see Chapter 4 if you need a refresher).

Note: Just because an issue falls into the bottom-right-hand corner does not mean it does not need to be addressed – this process simply prioritizes the

FIGURE 7.1 Impact vs likeliness matrix

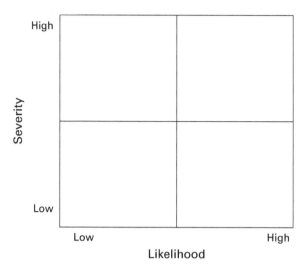

issues. Make sure you understand each issue fully and create a mitigation or monitoring solution in order to assess if issues change over time. Remember – TBD is about long-term thinking, not just short-term iceberg spotting.

Now you have the final quadrants mapped out, stand back and look at the issues that will most affect your business in the short term and long term.

TOP TIP: Pause at this stage and think about what these say about you as a business, your corporate culture and the future of your business – its direction, ethos and consumer trust. How will changing the issues you've highlighted affect these (and other) issues? How will you make sure only positive outcomes occur?

Step 6: Context (20 minutes). Finally, your killer issues need some context. How easily can competitors implement these issues? Draw out another matrix with ease of implementation (Easy to Hard) vertically and the severity of the impact on the business if it happened (Low to High) on the horizontal axis.

Step 7: Now kill the competitors (30 minutes). Once the most popular ideas have been identified, use the remaining time to brainstorm ways in which these issues can be neutralized, lessened or mitigated against. In other words, how can they stop the competition now that they know what isn't working and where the potential areas of weakness are? You may want to prime the group slightly by using words and phrases like:

FIGURE 7.2 Implementation vs impact matrix

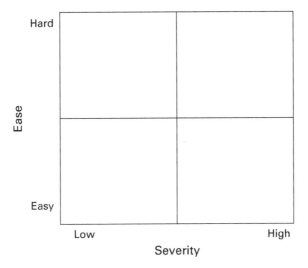

- What strategies can we...
- How do we...
- What do you recommend that we...
- Stopping them...
- In order to differentiate us...
- How could we protect ourselves from...

These words and phrases focus people on the task and the group at the same time – a key moment of focus that can bind people together towards achieving a common goal.

Write down each solution next to the corresponding idea. Do not rank them or ask for more clarification unless the idea or implementation is totally unclear – it is ok not to have all the details at this stage.

Step 8: Relay the findings and demonstrate clear next steps (10 minutes). Once all the ideas are exhausted or the room feels like it's coming to a good point to break, review the whole activity. Thank the participants again for their time and candour. Make them aware of the next steps in the process; these will be different for each company but usually there will be some form of review. This is a good time to ask if there are one or two people who would like to volunteer to keep the group informed (with the help of yourself/whoever else should be involved). This is not necessarily vital but it is a good way to keep people involved and make sure hierarchy mentality is avoided where it is unnecessary. Spend as much time on this step as you need based on your corporate culture – this is a key moment you have to 'sell' change into the wider business.

TOP TIP: This step is a good point to congratulate, thank and identify good behaviours. Sometimes it can be a fun task to award funny faux awards like 'Most out-of-the-box idea', 'Most likely to...' etc.

The benefits of this activity are far-reaching – not only does it free up employees to think about big and small ideas that have an impact on the business, but the ensuing discussion also instils a sense of pride and commitment when ideas are identified and acknowledged.

Sometimes you just need one thing... someone else

When clients ask me to talk on the subjects of change, disruption and emerging technology, I am often met with warmth from senior levels that (I am told) the client does not see across the organization. Clients often talk to me about how frustrating this is, to which I usually remark, 'You are not alone – lots of companies have the same behaviour... it's not necessarily bad' or something similar. The company (but really senior decision makers) simply needs to hear someone else talk about the issues that have been raised or documented before, in order to check a box and reduce the risk of being wrong. If this scenario sounds familiar to you, the next few chapters will really be of help when coming up with your future strategies.

Often, but certainly not always, the best results are seen when companies utilize the skills of outside experts, industry figureheads, business leaders or consultants. Using such individuals not only makes the process look like the company takes the desired outcome seriously but also means your company gains multiple viewpoints to the proceedings. Some CEOs tell me they feel uncomfortable bringing in outside help. When I hear this I tell them about Sun Microsystem CEO Bill Joy's famous quote: 'No matter who you are, most of the smartest people work for someone else.'

Using outside counsel when it comes to disruptive technologies

You already have a good system with TBD+ but you can get extra insights or replace personnel with experts if need be. In this instance, it is a good idea to focus the efforts of the outsider in order to maximize the benefit to the company. Here are some tips for working with external experts and consultants:

- Focus the brief – make sure you are clear on what you want the individual or team to achieve. This could be general coaching, planning or perhaps an oversight function – each would require a very different brief from you.

- Think about additional introductions – using the Expert Referral Staircase methodology (see Chapter 5), you can get a lot of value from external parties by leveraging their networks via introductions and contacts. (Note: This process works best when it isn't contractual and beware of those who fail to keep promises in this arena.)

- Add on training – many businesses use experts and consultants on a single project and forget about the individuals who then have to bring it home. The best results I have seen are from teams who have a clear goal and then create opportunities for additional development and practical advice on how to implement what is required of them.

Conclusion

As we've seen already, disruptive technologies and innovations scare people because of the speed at which they work and the impact they have on businesses. Changing this mentality is tough because of historical evidence, personal reasons for maintaining the status quo and/or lack of knowledge of impending events. However, change isn't impossible and thanks to COVID-19, you'll never have a better time/excuse/reason to change what's not working in your business (and life).

Clients tell me that change is increasingly hard as the pace of change accelerates within and outside their organizations. There's nothing easy about change but you can change one thing easily – your attitude (to change). There are often big opportunities attached for businesses who go beyond simply avoiding icebergs. Ask yourself a question: on a scale of 0–10, zero being 'not successful at all' and 10 being 'I could not be more successful', how successful over the past year have you been at pushing new ideas forward? Or put another way, how many approvals have you gained over the past year for big-ticket items? If you're not answering six or above, the next chapters will help you increase this ability by helping you to spot potential issues and create a greater chance of success.

This chapter explored some extremes of ongoing movement – a push towards being more open, transparent and accountable in order to create the best products and services and have a happy workforce. In the following chapter, we will specifically look at the issues you can expect to encounter and should be on the lookout for.

08

What to look out for

This chapter is about roadblocks, speed bumps and problems that will occur on your journey.

Specifically, you will learn to:

- look at why disruptive strategies can go wrong (and what to do about it);
- understand what to do when disruption threatens your company;
- understand the icebergs and how to navigate the different types;
- persuade your company to accept change;
- explore some tried and tested ways to encourage disruption, innovation and innovative thinking.

As previous chapters have described, disruption happens for a wide number of reasons but generally when things aren't quite right or a company gets complacent with the status quo. Numerous companies continue to fail to change for a wide variety of reasons: 'big ships need time to turn', the company refuses to see it coming, 'our people won't do it', and other responses are common. TBD is a great way to avoid these scenarios but even when you use TBD, surprises and problems can and will arise. This chapter covers what to do when these (and the other) issues arise.

Expect the unexpected – they will come

Entertainment, transport, music, media and manufacturing – all have gone, or are going, through massive changes due to technology and shifting consumer expectations. Having spoken to companies that have gone out of business or that are being hit side-on by technologies like 3D printing, many just didn't see the mythical 'it' coming. I say mythical because 'it' is rarely one thing but rather a collection of events or elements that 'conspired' against them. Rightly or wrongly, this attitude helps no one and because the companies did not expect disruption, when it happened the effects were that much worse.

Disruption often comes from the side or the outskirts of industries – the little guy, the start-up or the guy who people said 'no' to. Having looked at these people I see they have things in common: they expect change, they make it happen and they seek out difference. Most companies do the opposite – does yours? If instead you expect disruption, you can plan for it and lessen the impact, potentially staving it off completely or even turning it to your advantage. Some clients I have worked with have taken this thinking to another level – effectively assuming that new models will supersede their current business model and creating scenario plans for this eventuality. Not only did this create new ideas and highlight key issues but it also enabled them to identify multiple routes or options open to them. Instead of fearing change, they chose to be proactive and face the issues before they could become issues. These companies see disruption as a part of business, not as one-off occurrences. Thinking like this is not puffery or naivety; it is a strategy to reframe a situation and gain the most from it. The key here is the mindset towards disruption, not necessarily what you do when it arrives; what you do before it arrives often determines the ultimate success or failure of the disruptive force.

'Dematurity' is a great term that has been created for old industries that are becoming, will become, or are ripe for disruption because of new technologies like artificial intelligence, 3D printing and nanotechnology. Whether you are in a new or an old industry, the ways of reaching consumers are changing, technology platforms are shifting year upon year, and new behaviours are increasing the demand for new products as well as innovation from existing ones at an unprecedented pace. TBD has demonstrated how some of these changes can be identified, foreseen and turned to your advantage but if history tells us anything it is that not everything can be predicted. This is why a flexible approach like TBD is the best strategy and why flexible, open businesses are often the most resilient to outside forces.

Remember, there are always multiple options open to you and not all disruption is bad or has negative effects. You may identify new partners, smarter ways of doing things, additional products that can help people (and create revenue and jobs) or you might simply replicate the new idea in your own way. As business becomes more diverse and less predictable, an ability to roll with (and learn from) the punches may just be the best skill you ever learn.

Run away if you hear senior executives say 'it won't happen to us'

Prediction is hard at the best of times (see earlier chapters for examples) but business predictions are often even harder because different biases act on us and the pressures are different than for, say, predicting what 3D printing means for us personally. The TBD framework is built on using what you know versus what you think may happen; however, disruption can still blindside companies. It is important to keep a keen eye on outliers and maverick companies using the TBD methodology, as well as completing a quick quarterly sweep along with a robust monitoring and alert strategy. Unfortunately, some industries will feel the effects or stings of disruption faster (and more acutely) than others because of their interdependence or traditional structures. The following is a guide to these industries with examples in brackets of technologies, platforms and people attempting to give them sleepless nights.

Industries that have already felt the effects of disruption:

- auto (new competitors, lower ownership desire, ride-sharing and Uber);
- retail (Amazon, online price checking);
- TV/media (internet, Netflix, Facebook, Google News, augmented reality, virtual reality, Snapchat);
- telecom (VOIP, messengers, challenger brands, Google);
- insurance (automation, standardization, peer-to-peer networks, telematics, price-comparison websites).

Industries that are feeling the effects of disruption right now:

- finance (new challenger brands, payment technologies, blockchain technology, removing profit centres);
- manufacturing (3D printing, distribution, on-demand, new longer-lasting materials);
- telecom (messengers, Google Fiber, VOIP);
- grocery (Amazon, Deliveroo, food box delivery);
- construction (3D printing, graphene, nanotechnology).

Industries that are about to feel the effects of disruption:

- healthcare (3D printing, connectedness, social networking, big data, AI);

- education (social networks, Lynda, MOOCs);
- law (peer-to-peer networks, on-demand economy, AI);
- HR (AI, machine learning, automated processing);
- wealth management (AI, machine learning, cryptocurrencies).

You should ask yourself the following: Is your industry listed above? Is it where you think it should be? If it isn't listed, what bucket would you put it in? Why? What does this tell you?

Remember: fearing the iceberg is wrong

Often forming naturally or breaking away from larger ice formations, icebergs can be beautiful, dangerous and signals of much larger issues. The danger comes from the 80–90 per cent of the ice below the surface of the water – this is probably why they are often used as metaphors in explanations of disruption and disruptive or emerging technology.

Instead of a fear or warning metaphor, I find it useful to use the iceberg analogy to help innovation or disruption 'champions'. Whatever term you want to use, the important thing is to understand the challenge ahead and what will be required in order to change a business mindset or the direction a company is going in.

Draw an iceberg in your head. Now imagine you have sliced through it vertically so you can see a cross-section. You can see 10–20 per cent above the water and 80–90 per cent will be below the water. Above the water is what success looks like. This is what people see. Think about what would be on this part of the iceberg. It might be increased profits, it might be more staff, or it might be a new product. Whatever it is, visualize it there.

Now think about the 80–90 per cent underneath the water. This is your job. This is what people don't see. This is the hard work part. This is the segment that will take your dedication and your commitment. It will require good habits and cause you frustrations. List these out now on a sheet of paper – what will frustrate you along the way or during this process? Be specific – name names. Be ruthless. Now put this piece of paper somewhere safe. Think about the disappointments you may have along the way and the sacrifices you will need to make in order to push this 'over the line'. Think about what will go wrong. What will fail and what will be in your way? What are you going to have to persist through? Write these things down too.

Now you have two lists of negative things. Are these insurmountable? Do they look smaller now you've written them down? Does it feel like an action list waiting to happen? Could you mitigate anything straightaway? Should you?

This exercise is a good one for visualizing issues before they arise – they can form part of a mitigation strategy session or simply a star to chart your course by and check in on from time to time. Not everything on your list will happen but knowing about it allows you to think about the effects it could have and what (if anything) you want to do about it. Unfortunately, when it comes to big changes, you rarely get the glory without the guts. Understanding this is key in the war to win over hearts and minds – other people's and your own. Stay the course, knowing that doing so will be worth it.

What you should do and what you shouldn't do

Experience tells me that three core things happen when disruption is expected and companies begin to plan for it. Some are good, some are less good – here are some core do's and don'ts.

1 A transformational individual is brought in to 'save the day'

Whether it is Chief Transformation Officer, Head of Digital Transformation, Chief Change Officer, Innovation Director, Disruption Specialist, Chief Innovation Technologist, Chief Growth Officer, Innovation Hacker, Technology Hackologist – the name is irrelevant; one person is often brought in or promoted to fix the issues and act as the face of the fight. There's nothing particularly wrong with this approach – it has worked before. Some who have been put in this difficult job have seen great success, but the vast majority see frustration and often quit one or two years into the job when little has been achieved.

DO: FOCUS AFTER A SOLID INITIAL REVIEW
If you must focus on an individual, review the deliverables carefully. Often the bar is set way too high before the individual or company considers all issues relevant to the bar being set.

DON'T: IGNORE THE REST OF THE STAFF
Hiring a big-name, high-profile or outside consultant is a tried and tested way of bringing in talent that won't go away overnight. However, be

cautious of those who do not make empowerment of the existing staff to help them (and the company) the first issue they address upon arrival. Some corporate cultures can sustain this 'impact', but others can be polarized (or worse, damaged) by the 'individual' fixer coming in. Audit the staff, plug weaknesses and create a regular opportunity to increase skills – the majority of staff will be interested.

DON'T: HIDE BEHIND A TITLE

Understanding a job title is core to how people approach and think about the individual – having a title that is easy to understand is a smart move to focus everyone on how this is part of business as usual despite the fact you are trying new things. Communication, not titles, is key.

2 The company lacks actual commitment

Commitment is a core reason why initiatives – big or small – fail. The ability to push things 'over the line', knock things 'out of the park' or simply deliver on promises is the difference between a company that is moving forward and a company that is in neutral, or worse, moving backwards. The ability to foster commitment is not easy and takes time. Once you have commitment, the hardest task is to keep that commitment.

DO: TALK ABOUT WHAT WILL HAPPEN IF YOU LOSE COMMITMENT

The advice I give clients is to question commitment at all stages and make sure the other parties (often senior decision makers but not always) know what you will do if they fail by discussing it before it happens. Ramifications, knock-on effects and consequences are all good motivators. Setting expectations early, along with suitable predictions, enables you to look prepared, seasoned and in control. Not only this but you will feel more in control and be able to spot issues earlier.

DON'T: GET FRUSTRATED AND BECOME DEMOTIVATED

This is the easiest path but the least righteous one to take. Instead, investigate the why, where, when, how and who of the situation and see what you can fix. Remember, you are at least 80–90 per cent of the success of anything – buckle down, refocus and push forward.

3 Our/the business just doesn't have the right _____ to change

Attitude, understanding, mentality, knowledge and experience are all words clients have used to fill in the blank but really what they refer to is a poor corporate culture or environment in general. Companies these days have a hard time motivating staff for a number of reasons but change is a big area as it can cause uncertainty and fear in individuals of all ages. The current climate notwithstanding, many are in jobs they aren't fully happy with and so corporate culture activities often leave these people deflated or wanting. Culture is a hard thing to change but it can be done – changing it takes more of your 80–90 per cent.

DO: MAKE A SHORT-TERM STATEMENT AND A LONG-TERM STANCE

Whether it is something big or small, announce it. Many companies scrimp on the communication of initiatives but doing this undermines the impact. How you announce something has a huge impact on the way that thing is perceived and carried forward. Beyond this, the reinforcement of ideas and new initiatives can also be skimped on, so plan a long campaign, not a week of activities or a poster campaign, if you have something that is fundamental to your business; spend some time and spend some money – short- and especially long-term benefits will be more visible.

DON'T: IGNORE INFLUENTIAL INDIVIDUALS

While big 'shows' are good and necessary, most cultural change will come from people demonstrating and reinforcing good behaviours that others then follow and learn from. Not all people or businesses follow a shepherd and sheep metaphor but every company has individuals that others look up to or look to in order to gain a sense of where to go and what to do next. Use this to your advantage and subtly identify a list of individuals who are problem solvers, relied upon by others, who leverage relationships carefully, assess the political landscape astutely, keep others' confidences and are often long-time employees.

TOP TIP

Utilize key areas – transform main entrances on launch day, get everybody together, take photos, sell the ideas back, make them individually accountable by putting their faces against the idea of success. One company I worked with created a video wall that showed the progress of the task every day and used RFID (radio-frequency identification) tags in the employee badges to personalize the messages to each employee as they swiped in and out during non-peak times. This saw an increase in accountability, recall and goals achieved by individuals and groups in the organization.

Expect that things will also go wrong

By now you should be feeling a bit better about the uncertain future that lies ahead of you. You are aware of and understand several technologies that will disrupt multiple markets (3D printing, nanotechnology, artificial intelligence), you have a decision framework to help you evaluate issues (TBD) and a methodology for proactive ventures (TBD+). Things could be worse, right? Even with all these tools there are some common issues that can arise when the TBD process is used that I have observed over the years or that clients have relayed back to me. Here is a list of the most common issues and what to do about them.

Problem 1: You never repeat TBD after the initial session

TBD takes a significant amount of time the first time around; after this, things do get a lot faster and people set up alerts and calendar reminders to help them with an always-on approach.

SOLUTION

Blow up the grenade before it blows up on you. If you believe commitment to be a real issue, then you have to call it out to the relevant parties before it is a problem. Testing commitment is hard but there are some helpful phrases that can help you and other parties:

- Is there anything that would stop you from…?
- What would you like me to do if this happens?
- Experience tells me that…
- Is that going to be a problem? Want to talk about that?

TOP TIP

Using a freelancer to complete a quarterly update is not a crime. Everyone gets busy and resources are finite. While taking the time to do it yourself will usually yield richer results in both the short and long term, a solid brief and review process should mean you can be made aware of potential threats rather than forgetting or ignoring the process for a quarter.

Problem 2: Senior execs just don't get 'it'/see 'it'

Disruptive technologies require a level of buy-in and often the process can feel arbitrary if you aren't sensing anything snapping at your heels. This is the best time to complete the process; as Finley Peter Dunne once said about the job of a newspaper (Shedden, 2014), your job 'is to comfort the afflicted, and afflict the comforted'. Don't wait for your company to be in pain: pre-empt it.

SOLUTION

This can be a tough one. Moving people from negative or positive can be easy but neutral is the toughest position from which to create movement. Before you get downhearted, bring them on board early and ask them how much problems would cost to fix, how they have previously handled them and what it meant for them personally (longer working hours, stress, ill-health, unhappiness).

Problem 3: You get sideswiped by a technology nobody saw coming

Disruption comes in many forms and often does not follow a direct path – companies like Uber, Deliveroo, Tesla, Airbnb and Snapchat are great examples of this. Most were seen as annoyances at first by established players and forces before they dominated the headlines and upset multiple markets. Many CEOs and competitors look back on disruptive forces and remark that they discounted the technology too early. This element is why TBD builds in a review and mitigation layer for your business to keep a watchful eye on seemingly innocuous technologies that may later pose a threat (or opportunity). The key is being prepared.

SOLUTION

Run or re-run the 'kill the company' exercise from the previous chapter – focus on technologies specifically so future issues can be potentially spotted or areas further researched to avoid, neutralize or mitigate similar impacts. No company is immune from disruption – thinking this simply allows disruption to happen faster when it eventually occurs because there is no preparation or flexibility against the impending force.

Problem 4: You lose momentum after completing the initial TBD process

This is a common occurrence with disruptive technology projects because often, after the 'wow' moments and the planning, little immediately happens that is visual or 'wow' in nature. There is a lot of research and next steps. You have already seen how to claw back enough time to complete the TBD process but beyond this, maintaining enthusiasm and keeping people up to date can be hard and requires more time and willpower from you. Beyond this, people become busy and time does not expand.

SOLUTION

Plan for the lull. Often clients tell me this issue is a matter of perception rather than reality. You have lots of options: run a mini-TBD, call a quick meeting, use mini-chats or create a robust visual update for the entire company somewhere prominent (not their inboxes!). The choice is yours.

Problem 5: Your company isn't innovative enough and can't afford big agencies to help

TBD is run by agencies as much as it is by companies. Both have high and low points and different levels of expertise but neither is better or worse than the other at the process – some are just more open to it. Neither the agencies nor brands have infinite budgets or time to waste but the ones that get the most out of it understand the value TBD provides and what they are looking to achieve. Many of the businesses I have worked with or for stifle innovation by requiring the movement of ideas up through levels of management before getting any investment – sometimes this is intentional, other times often unknowing. It is important to think about whether this is something your business does before starting TBD as you may need to approach things slightly differently as a result.

SOLUTION

Creativity isn't the bastion of any one person or department – even creative agencies tell the world this at festivals like Cannes or conferences such as TED. Instead, the job is to shape, direct and lead. Good ideas can come from anywhere. Adobe knows this and has enabled its employees by creating the 'Kickbox', a kit the company gives to employees who have an idea (explained below). Your company may want to create a similar toolkit.

An innovation toolkit

The previous chapter talked about creating new mindsets and corporate behaviours – this isn't easy and results can be slow. Sometimes a large initiative is needed that, if promoted well, can really kickstart new behaviours and a surge of enthusiasm that is hard to dissipate. One of the best examples of this was the 'Kickbox' that Adobe created for its staff. Kickbox is a physical box with a big idea behind it; a senior executive of Adobe said he came up with the idea because he didn't like the odds of making large bets. Instead of investing (or risking) $1 million on an idea, by enabling 1,000 Kickboxes, Adobe can make 1,000 smaller bets that could pay off big in the short or long term – much better odds. According to Adobe's Kickbox website (kickbox.adobe.com) the kit is 'designed to increase innovator effectiveness, accelerate innovation velocity, and measurably improve innovation outcomes'.

The physical box contains a variety of items:

- instruction cards (known as 'levels');
- a pen;
- a timer;
- two Post-it note pads;
- two notebooks;
- a bar of chocolate;
- $10 Starbucks card;
- $1,000 pre-paid credit card.

You read that correctly, $1,000 on a credit card that the holder can use any way the box owner pleases without having to justify, ask for permission or worry about repercussions. By taking this route instead of the usual 'just expense it' route, Adobe avoids frustrating paperwork and approval friction, and offers a huge level of trust – almost guilt-inducing trust you might say. No one wants to waste money, right? After the card is depleted, it sends a signal to the user to pitch their idea to the senior team.

Note: A promise of money for ideas to be tested is another route but the trust element of the credit card method is the key ingredient here. By giving something immediately, Adobe fosters what psychologists call the Reciprocity Norm or Principle – if someone gives us something we feel compelled to give them something back of similar or greater value.

The process is a detailed procedure for innovative thinking and product or service development; there are six 'levels' and each has to be completed before moving on to the next. Before we explore the six levels it is important to note a few things:

- an employee at any level can request a box and their manager cannot veto it;
- the card can be used on anything the employee needs or would like, without ever having to justify it or fill out an expense report;
- the idea gives the individual permission rather than approval.

The levels are an important part of the process and are not meant to stifle creativity but instead guide users through the process to maximize their chance for success:

Level 1: Inception. This level talks about purpose, motivations and the route to success.

Level 2: Ideate. Helps the box holder learn methods to help creativity flow by focusing on how the world should be and not how it is.

Level 3: Improve. This level helps the box holder understand that not all ideas are good and how to determine what makes a good idea.

Level 4: Investigate. This level helps the user evaluate an idea with simple experiments.

Level 5: Iterate. This level helps the user understand the data from Level 4 and move the idea or ideas on to discover 'the true nature of your idea'.

Level 6: Infiltrate. The final level is about 'corporate combat' where the user is helped to create a winning pitch using data the organization wants to see.

As senior execs point out when interviewed about the product, Adobe only needs one bet to pay off to pay for the entire project. Since distributing more than 1,000 of the boxes, Adobe reports that more than 20 completed boxes have been 'sold' to senior management – that's more than 20 ideas that were previously locked away or never thought of. One of the ideas even led to a more than $800 million acquisition of stock photography company Fotolia, which cemented a larger product (Adobe's Creative Cloud) marketplace position and access to a different part of the worldwide designer audience.

> **TOP TIP**
>
> Create your own version of the kit – Adobe allows you to do this and encourages using the Kickbox in your organization. For more information or to download your own version of Adobe's Kickbox head to: kickbox.adobe.com.

Conclusion

This chapter focused on changing the perception that disruption is unlikely into one of it as inevitable, and that it makes good business sense to prepare as much as possible for this so the impact will be as small as possible. Getting the right people involved is key and smashing preconceptions as well as opening up is imperative for businesses who are looking to really step away from the crowd and not just survive. The next chapter focuses on additional elements your company needs in order to make it to 2026 and beyond with the least amount of disruption possible.

09

Dis-innovation

This chapter focuses on larger topics that will impact the future of your business and disruptive technology strategies:

- why just thinking differently won't save you;
- why thinking 'big changes are hard' and 'small changes are easy' is wrong;
- what needs to change in order for your company to be ready for 2026.

So far in the book we have discussed and touched upon some key indicators surrounding disruption: distribution changes, production efficiencies, consumer behaviour shifts, competition coming from the side and new laws or regulations among others. Each of these changes has implications and responses that will differ based on the industry you are in and the type of culture you have. The key – beyond spotting them early – is reacting accordingly. In the last chapter, we spotted icebergs and looked at how to respond to stalls, poor corporate culture and other issues. This chapter will focus on your company thinking moving forward, specifically around 'thinking differently'.

Why 'think different' probably won't work for you but thinking differently will

Apple famously used the slogan 'Think Different' in a 1997 ad campaign created by TBWA/Chiat/Day in response to IBM's simple 'Think' slogan. The former slogan has since gone down in history as the epitome of how businesses should think about everything but it is very wrong for most businesses.

'Think Different' works for Apple because they are Apple and fundamentally it is what the company stands for about everything. Apple moves beyond technology and looks more at the hand and mind that is to use the technology. As one of the biggest and most frequently recognized brands currently out there – and for the foreseeable future – thinking differently works for them because it is synonymous with their identity. Can the same be said for other businesses? Probably not. If you haven't already, take a moment to review Simon Sinek's TED talk about Apple (https://www.ted.com/talks/simon_sinek_how_great_leaders_inspire_action?) and you'll see why Apple does what Apple does. The key is understanding your 'why'. The aim of most businesses isn't to be Apple or even similar to Apple, but to learn from their behaviours to create ones that work in and around your business.

Success rarely lies around simply copying something that worked for someone else; the best way forward is adapting it and creating the perfect fit for your needs. Therefore, this chapter will help you to actually think differently rather than simply telling you that Apple's doctrine is the way forward for your company.

How you fix a bench says a lot about you and your company

Disruption is and comes from everywhere – it is a constant in our world. Without it we would move forward slower and create fewer truly new things able to propel industries, economies and such. However, most companies would rather continue to fix surface issues in the hope that these will fix the underlying issues, problems or rot. I call this bench theory.

Some people use a bench the 'normal' way for its entire life – that is to say as a table to sit at. Others choose to use that bench as a workstation, creating things on it. Others simply use the bench as a place to store things. In reality, there is no right or wrong way to use a bench. However, when that bench has a mark on it or a crack appears, how you react and make your next decision is key. Do you:

- Apply a layer of lacquer?
- Cut out the bad part and replace it?
- Throw the bench away and get another one – a better bench?
- Give up on the bench and use the good wood for something else?
- Do nothing – it's just a crack.

There is no right answer but the answer you pick says volumes about you and the way your business reacts to issues. Look at which answer you picked and explore that on a blank piece of paper. Put the decision in the middle of the paper and create a spider diagram answering questions like:

- Why did I decide to _____?
- What elements lead me to believe that this particular course of action will occur?
- What does this say about the way I approach things that are broken or need mending?
- What should I be wary of in the future?

Many people make decisions in their everyday lives, and increasingly work-related ones, out of fear concealed as practicality; this is also known as the 'easy' route, the 'right now it's fine' route and the 'don't make waves' route. I am telling you now – make waves because things are not fine. If you think things are fine, re-read the Introduction to this book. Change is coming faster and faster.

Complacency is the killer, commitment is key

Often you know the right course of action (or the desired outcome) but the first practical steps seem so huge or insurmountable that you don't change or you implement a non-change (something so small it is barely noticeable and has no effect other than ticking a box). Often what is asked of us or what we want to happen seems so 'out there', 'impossible' or 'not us' that people don't even try, and this is deadly to a business. Those who do ask for it often actually make it happen. I am telling you now, ask for it. Those that do, with the right support, attitude and conviction, make things happen. You don't have to be Steve Jobs or anyone else; just remain aware and focused. You have permission to make it happen. Be hell-bent on pushing things forward and making a difference. People who do this – even if whatever 'it' is doesn't happen right away – are on the whole happier, more satisfied and more likely to inspire others around them. At the very least they are fighting for something they want and not waiting for something to impact them. Take the chance. Few jobs are forever – this is especially true for the younger generations coming into the workforce (see the next chapter) – so make every move count.

The trouble with thinking differently is that the act itself is often only committed to for a short period of time; a brainstorm, an away day, a pilot or a short test that yields no results and therefore the whole idea is abandoned. I see these issues arise time and again but especially with disruptive technologies because of the unknown factor, individual priorities, the immediate value and, sadly, greed. You can stem these issues, though, and it's not even difficult, but it does take a bit of consideration.

Your approach to any change has to be tailored

There are vast differences between helping financial clients with change and disruption and helping agencies or car manufacturers. Each industry is unique and while there is a lot of crossover, every company in and of itself is a different beast. However, the same processes, exercises and tools can be used to understand the issues at hand.

As previously mentioned, 'think different' isn't for everyone (or indeed anyone other than Apple) but thinking differently is. This is not giving you permission to stop thinking in a big and complex way or saying that small changes can't make a difference but it should inspire you to do these things in new ways. The 'old ways' fail or often fail to give us the best (or freshest) solutions to increasingly complex problems. All too often, clients tell me 'the devil's advocate' comes out to play and shoots down good ideas with clever criticism rather than addressing realistic issues.

Tom Kelley, CEO of IDEO – a design and innovation consulting firm – believes devil's advocates are a huge problem in the business world and discusses how to handle them in his second book, *The Ten Faces of Innovation* (2008). Kelley believes devil's advocates smother 'fragile' ideas with negativity and so wanted to stop this behaviour happening so frequently. He identifies 10 types of people that businesses have in different combinations and shows, through examples, how each helps different businesses evolve, adapt and diffuse change and disruptive forces.

Considering your business's personas is important before launching anything – beyond simply making more sense for tonal and imagery considerations, understanding what drives and inspires your workforce is a smart way to roll out new initiatives and ideas. See Kelley's 10 personas below.

'The Learning Personas' (no one can be complacent):

1   The Anthropologist – observes behaviour and brings new skills into the business through deep understanding of how people behave and interact with each other and products or services.

2   The Experimenter – regularly comes up with new prototypes and is a key proponent of trial and error.

3   The Cross-Pollinator – looks at other cultures and industries and creates fits for your organization. Described as analogous thinking, this approach is explored further towards the end of this chapter.

'The Organizing Personas' (ideas are games of chess – everything must be considered):

4   The Hurdler – understands how to overcome obstacles within a company to achieve innovation.

5   The Collaborator – brings disparate groups together to achieve unique and considered new solutions.

6   The Director – beyond bringing the right people together, this persona inspires the assembled team to achieve more than would be possible if they were alone.

'The Building Personas' (ideas happen when people are empowered and are channelled):

7   The Experience Architect – focuses on needs, the experience and how people connect with the final result.

8   The Set Designer – helps other people do their best by creating the right environments around people.

9   The Caregiver – service as standard is the start for this persona – they anticipate needs and create to serve them.

10 The Storyteller – inspires action by tapping into fundamental values, beliefs and traits.

Before you go on, think about your company. What are you made up of? Ten per cent Caregivers, 90 per cent Set Designers? Perhaps the reverse? What does this mean for future programme rollouts? What does it mean for your hiring policies? Training requirements? Beyond your initial gut reaction, how are you going to challenge these assumptions to know the real mix of personas that make up your company?

TOP TIP

The devil's advocate has a reason for being there; sometimes the devil may be right. It is important to listen to all concerns but to think positively and openly. Dismissing concerns is also not good for individual or group morale. Instead, as Kelley suggests, encourage your innovation persona by saying, 'Let me be a Collaborator for a moment; our customers believe in good value and the environment, so we should make sure we talk to buying and PR before we decide anything.'

TOP TIP

Think long and hard about how you will communicate 'thinking differently' projects. In my work with different industries, one thing has become clear; people like new opportunities but fear the unknown. When I work with clients I often get funny looks when I start to ask about communal areas and how this will be rolled out, as if success is a given. Instead I urge clients (and you) to think about and assess the corporate culture, current general mood and tone. These three elements are imperative for success and increased favourability towards future projects and getting budgets approved. Often new projects fail before they begin because of the way they are introduced. Don't skip thinking about the introduction.

Remember Bill Joy's quote in Chapter 7? 'No matter who you are, most of the smartest people work for someone else.' Now imagine an extra 3 billion new minds out there – this is the number of people coming online in the next 10 years (or less if current adoption rates increase). Many of these people may never be full-time employees of any corporation and many will have learnt their skills and work exclusively online. Let that hit you for a minute. That's a lot of new people to add to the other ones already out there. People will have different viewpoints, experiences, networks, resources and ideas. Imagine the possibilities of bringing that sort of vision to fruition to help your company.

Thinking differently requires other people, not just you

Disruptive technologies are often – but not always – totally new. In other words, they are usually the great unknown, which means deducing how they will impact you or how your rollout of the technology will be difficult. Deduction is often not possible despite many companies running complex modelling and looking at multiple sources of data. Instead, companies often have no choice but to effectively commit to trial and error. The best option for new products or services created from disruptive technologies and forces may just be to launch and learn.

Analogous thinking is a solution to the issues above. The emerging practice is becoming more and more popular as the benefits are seen throughout organizations and departments (and not just start-ups; small–medium enterprises and big businesses are using this method). Analogy and 'analogical reasoning' has been used for many years in fields like biology, adaptive systems, software and computing to aid understanding and solve problems but more recently it has evolved into the world of business as 'analogous thinking'. Analogous thinking is the process of looking for help with known issues by referring to and connecting with unknown but potentially similar areas with the aim of transferring and adapting learnings from one to the other. The reason for the increase in interest is twofold: first, analogies are everywhere and somewhat easy to mould but second, and more importantly, analogies enable people (leaders, strategists and such) to look at the picture more widely and evaluate decisions in a new way or through a new perspective. As we have seen via the TBD framework, understanding how decisions are made is crucial. Analogy can help decision makers make more informed choices – especially when faced with emerging and disruptive technologies (often unknown entities), as solutions come through applying the same or similar lessons from previous experiences or from something the decision maker heard about. Unfortunately, few people actually realize they make decisions in this way and this is a huge problem.

Bad analogies are of course rife throughout the business world. Going beyond bad analogies and challenging yourself to stop using them is step 1. Step 2 is to challenge your own tendencies to seek out information that confirms your (own and your company's) beliefs. Psychologists refer to this as the 'confirmation bias'. Combating this bias is tough but it is possible. First, you must look at the sources you are using and how they don't relate to the issue you have. It sounds counterintuitive but you must try to make the analogy not fit. Only by doing this will you really use quality inputs for your solution and arrive at a quality output. When looking for differences it is unlikely that you won't find any but if you don't this is totally fine. Your next job is to adjust for the differences that you have found so that you can then map or learn from the different thing, scenario or event. Once this has been done, you can really look at the solution you have in front of you and further adjust or finalize based on additional knowledge or elements. The key is to really focus on justifying why the item you are assessing accurately correlates to your issue (or why it doesn't). In doing this rigorously you will avoid creating cosmetic or weak analogies that will not help you in the long run.

Studies have shown that solutions provided by people from similar (but different) markets often have lower immediate usefulness but the amount of ingenuity, creativity and novelty is significantly higher compared to ideas generated by people from the same industry. In fact, more quality creative solutions are returned the further the field is from your own.

Chapter 5 introduced the Expert Referral Staircase (a process that helps you get increasingly rich information from experts recommended by other experts) which, when slightly adapted, can be used to identify areas where different experiences and analogies may exist for you to utilize to your business's advantage.

First, think about the issue you want solved – be as specific as possible. As you have read above, being specific at the beginning of any analogy-mapping process is key for success. Once you have the issue or the area you need help with, you need to identify the person you would initially ask for help if you alone had to solve the issue. For the greatest potential of success, having several ways to describe the same thing is often helpful here as it unlocks different connections in the brain. The third step is sufficiently framing the question or request for help in such a way as to unlock the expert's superior knowledge or insight to help lead you in the right direction.

Instead of asking your initial target for situations or industries that are similar to the problem or issue you are looking to solve, hold back. This isn't a one-question meeting. Instead get the person thinking about your issue with a good description and superfluous but related chat or information in order to get them accessing the right areas of their brain. Some people are good at having questions shotgunned at them but most are not, so warm them up before the big ask.

Then, when you believe it is a good time to strike, adapt or use one of these possible ways to frame your questioning:

- What do you think of when you think about issues to do with [CONDITION]?
- What two companies do you think would have similar problems?
- Who do you know who has had this or a similar problem?
- Who else would possibly suffer from [ISSUE]?
- Who else uses something similar to [PRODUCT]?

Depending on what you ask and the issue you have you may need to ask for the next person on the staircase – do this when the conversation comes to a natural end, then repeat until you believe you have reached the peak of the staircase and the ideas being received seem ridiculously implausible.

Once you have done this, do a little more research (including asking the person who gave you the information) to formulate your answers.

TOP TIP

Avoid email with this as much as possible – you may want to prime the person being asked beforehand but it is best to keep the purpose of the question a mystery as this often produces the best results. Remember, radical ideas can be found if you look in the right places – usually outside of your contacts. Use LinkedIn, Facebook Groups, community forums, university lecturers, app developers, anyone! Everyone can potentially help you with this task.

Based on a study by Franke, Poetz and Schreier in 2011, this method is a good starting point but the key is the tenacity in finding new people, not just friends, co-workers and the like – go outside your industry, go deep and go wide. For example, in the original study, a mask manufacturer was the client and the people who were asked included skateboarders, roofers and carpenters. An unlikely trio but each provided a unique angle on the same problem. You are looking for new answers so go wide.

Caution: Analogies are not foolproof. They require a lot of attention and many are often shoehorned into fitting a solution. Be wary of this and fight against it. Analogies are imperfect but can be revolutionary for companies who fight to find the right ones. Use swift methodological approaches and stop or refine what isn't working.

Anything that helps someone make a good decision should be applauded but often we unknowingly reason with ourselves based on previous experiences or because of external pressures. However, if you focus on how you make decisions, you can often make better ones and make fewer mistakes. Strategists are incredibly guilty of this subconscious reasoning, but while these folks tend to abhor being challenged, the best ones will always take it in their stride, challenge their assumptions and engage in analogous thinking. You have to accept the flaws and embrace them so you can turn seemingly mediocre ideas into ones that could save your business from disruption.

Conclusion

This chapter is a critical one to help you adopt the right mindset to push your strategy forward. Beyond this it is a realistic one that helps you predict and handle issues that you will come across (and therefore your apprehension should be lower as a result). Completing TBD is step 3 on a never-ending set of steps – making things happen from it is the hard part. Using analogies and maintaining a constant vigilance on the direction and the map you are using is a good strategy for success.

The next few chapters move away from strategy slightly more to discuss specific elements that are going to impact your thinking and decisions. First we're going to explore the next evolution of the web or 'Web3' as it's becoming known. After this we'll drill down specifically into the metaverse, a key area to really understand and think about as technology like VR improves and permeates more areas. Following this, we'll look at the human side of disruption with a specific focus on the young worker demographics coming into the workforce (often referred to as 'Millennials' and 'Gen Z'). First up, Web3....

The perils of writing a book about technology, especially disruptive and emerging technologies, is that areas often move both at blistering paces and glacially. Web 3.0, web3 or Web3 is one of the former. The following chapter takes a look at what web3 is, isn't and what it might just become.

10

Web 3.0: The opportunities... and issues

Before we explore Web3, let's talk about how we got here; it's important to understand the context. There's history, after all – the World Wide Web only turned 32 in 2021 – and knowing what came before is important.

First came Web1 (approximately 1990–2005) which – per Tim Berners Lee – was around openness, free protocols, community-focused and decentralized; there was little governance. Value came from the production, not the mechanics of how the network worked. Fifteen years on, Web 2 (more popularly known as 'Web 2.0') was coined. The shift was palpable; the open, free and, some said, hippyish nature of Web 1 got a wake-up call from Web 2.0. The focus now was centralized, owned, siloed entities that were controlled – in the main – by big corporations with interests more vested in profit and less about freedom and lack of governance. In reality, these entities (Meta, Google, Apple, Amazon) all utilized what the web offered and added a capitalism element, which worked and still works – but not without issues – today. The value of these networks is the difference between Web 1.0 and Web 2.0. In Web 2.0, the value is for the corporation and the information (often your information) fed into systems that algorithms then used to influence you further on that platform (and others). Sounds dystopian? You're not alone; the recent – and ongoing – big techlash isn't going away. The public has been abused for too long and seen too much for huge platforms to go untouched by government agencies. Both the European Union and the US government have multiple court cases and legislation pending in order to rein in the companies' monopolistic and anti-competitive behaviours. The outcomes will have ramifications for decades, not least of which will be Web 3.0.

Coined by Gavin Wood in 2014, Web 3.0 (or Web3 from here on out) is an umbrella term for a new chapter of the internet that has suffered – and to some extent is still suffering – from an early case of bad PR and buzzword-itis. The term has evolved to engender fear and excitement in equal measure thanks to booms, crashes and breaches. The latter is a large part of the issue – a lot of Web3 projects sound like pyramid or Ponzi schemes. Without significant and clear taxation, regulation and oversight, it was inevitable that nefarious forces would zoom into the technology as early adopters. The other issue is that Web3 isn't currently decentralized, from multiple perspectives. First, who owns what. The VCs have been pumping money into Web3 companies for years and both a16z and Pantera have established a monopoly within the space, with footholds in the majority of high-profile Web3 companies and often with shares which grant them 20× the voting power of regular shares. In other words, the VCs can vastly influence the major decisions companies will make down the line – the same VCs that funded, and still fund, a lot of Web2.

Putting the early issues to one side, as both governments and organizations are looking at these issues, Web3 offers, proponents say, a chance to take back some control from the big companies. The reason is simple: the data is stored everywhere, i.e. decentralized and without an owner. The idea that people will own their data is intriguing, for large companies as much as it is for the individual.

Why is Web3 disruptive?

Web3 is built on a central premise of – or you might say, conceptualized on – decentralization. The idea that ownership is with the users and not the large technology platforms is key. Identity is fluid between platforms, and data is – in essence – free. In order to do this, developers will need to create and utilize new protocols (rules) and infrastructure to break the control that platforms have previously had. Each application and each website or platform doesn't live in one place; little bits 'live' across lots of different devices (or nodes). The network is peer-to-peer and not client–server. You might have heard the phrase peer-to-peer when music file-sharing services like Napster came about, as they use similar technology.

Decentralized networks offer multiple advantages:

- Lack of central trust agent: no one has to know or trust anyone, and this can lead to more data being shared between partners.

- Disruption is harder: if one server crashes or gets hacked, the content, site – or whatever – stays up. In other words, you don't have to put trust in one place.

- They reduce points of weakness: censorship is less likely, as you have to shut a lot down for elements to disappear.

- Resources are optimized: thanks to the multiple nodes being used, no one server or resource is taxed too hard.

However, decentralized networks have downsides too, the main ones being nefarious users and use cases that have massive effects and implications. File-sharing is the largest and has many aspects to it, from legal content to illegal content. The anonymity that the decentralized networks allow often attracts criminals because they can move about freely, complete transactions and monetize illegal activities and subjects. Web3 has this potential and in some respects won't be able to guard against such things without breaking the fundamentals of decentralization.

Ownership is very much a focus for Web3 too. The ability to trace authenticity or provenance (as we saw in blockchain technologies in earlier chapters) is fundamental and therefore Web3 relies heavily on new acronyms like NFTs, DeFi, POAP, DAO, FOMO, Dapp, and more recent terms such as apes, moons, gas, stablecoins, multiverse, metaverse. Before we look at how people are using the decentralized web, or Web 3, let's make sure we understand what the main terms are talking about.

TOP TIP

Create your own blockchain 'degree'. Blockgeeks (founded in 2016) is an accessible community of over 250,000 'students' who are using the almost 2,000 courses on blockchain and related technologies like smart contracts to help them understand the world that is coming. From 101s to step-by-step guides, there's a starting point for everyone. Head over to blockgeeks.com to find out more.

A quick Web3 glossary

Crypto Shorthand for cryptography and cryptocurrency. The latter is any currency that exists virtually or digitally and utilizes cryptography to make transactions secure without a central issuing or regulatory authority. All transactions are recorded via a decentralized system.

Decentralized autonomous organization (DAO)   Most organizations have a hierarchy, with a central leader (CEO), but in a DAO the opposite is true, there's no central leadership. Instead of one person making the rules, the group agree to a set of rules that are enforced using blockchain technologies (smart contracts). Created in 2016 by multiple members of the Ethereum community, DAOs are not without their issues.

Decentralized finance (DeFi)   Umbrella term for emerging financial technologies that are based on secure distributed ledgers.

Decentralized application (dApp/dapps)   Refers to any website or application built on or using blockchain technology. An example could be a fitness application that rewards users for using the app with cryptocurrency.

Token   In the regular world, a token is something that represents something else. For example, a casino chip or a concert ticket. In computing, there are several types of tokens, but when it comes to Web3, a token is usually just another word for cryptocurrency or a crypto asset. Depending on the context, tokens have a wide range of possible functions from security to financial transactions to selling items in video games. Each can be traded or retained like any other cryptocurrency.

Non-fungible token (NFT)   Digital assets that are created, traded and verified using blockchain technology. Think of NFTs as akin to virtual baseball cards or any other collectable. Fungible assets like Bitcoin and Ethereum (the main two) all hold value. A non-fungible asset stores data like the creator's name, title and contact details. Owners of an NFT essentially own a created piece of work that they perceive as valuable and that could go up in value – both or neither of which could be true.

A popular, albeit somewhat controversial, example is the Bored Ape Yacht Club (aka Bored Ape), a collection of images of cartoon apes in various clothing, hats and expressions that is created by an algorithm. The group is now pushing beyond NFTs with other content, including films. Another popular example is the artist Beeple (aka Mike Winkelmann) whose first NFT, 'Everydays: The First 5000 Days', sold for $68 million at Christie's auction house in March 2021. The image was created by combining the totality of artworks Beeple had been creating every day since May 2007 and is a 21,069 × 21,069-pixel collage. Art is just one area that NFTs touch though; they can also be constructed or utilized in everything from gaming to profile pictures on Twitter and even owning things in virtual worlds. NFTs, while controversial, aren't without utility either; some have been used to grant the holder special access and privileges in the real world at events and online gatherings. Examples of marketplaces where art NFTs are sold include OpenSea, SuperRare and Foundation.

Decentralized exchange (DEX)   A website or application that enables users to buy, trade and exchange various crypto assets and tokens. Popular examples include Coinbase. Crypto.com, Gemini and BitMart.

Smart contract   Term to describe programs stored on a blockchain that begin when predetermined conditions are met and verified (think 'If X and Y happen, do Z'). Usually, this is to do with an agreement where all parties know the outcome but can also trigger other workflows or actions. No intermediary is needed, which saves time and money. Speed and trust (because no one else is involved) are big advantages for smart contracts.

Liquidity pool   A collection of cryptocurrencies or tokens that are locked together in a smart contract (a program stored on a blockchain that runs when predetermined conditions are met). A liquidity pool is akin to adding financial value to a game's points system.

Proof of Attendance Protocol (POAP)   A protocol that creates digital badges or collectables through blockchain technology.

Stablecoin   A digital currency that is tied to a 'stable' reserve asset like gold, the Euro or the US dollar. Stablecoins are designed to reduce volatility relative to unpegged cryptocurrencies like Bitcoin. There are three types of Stablecoin: fiat (backed by fiat currencies), crypto (backed by crypto assets and smart contracts) and algorithmic (supply and demand are algorithmically balanced). Examples of popular stable coins include Tether, DAI, Binance and USD Coin.

How people are using Web3 today

The heaviest utilization is finance. As is the case with most emerging technologies, it is, after all, the underpinning of most current value systems out there. People are using cryptocurrencies and their respective services (DAPs) for payment, lending, borrowing, crowdfunding, swaps, trading, insurance, and portfolio investments. Pretty much everything you'd expect with anything that has currency in its name. NFTs are probably the most well-known aspect of Web3 due to their meteoric rise and subsequent crash, which generates continual interest and media coverage.

Not everyone is convinced. Perhaps the most famous naysayer is Dan Olson, a famous YouTuber, who created the much-shared documentary 'The Line Goes Up'. In the documentary, Olson explores the dark underbelly of the NFT market which he likens to a poverty trap:

> The fact that tokens representing ape [profile picture collections] are useless, yet somehow still expensive, isn't an overlooked glitch in the system, it's half the point. It's a digital extension of inconvenient fashion. It's a brag, a flex and a form of mythmaking. And that's how it draws in the bottom: people who feel their opportunities shrinking, who see the system closing around them, who have become isolated by social media and a global pandemic, who feel the future getting smaller, and people pressured by the casualization of work as jobs are dissolved into the gig economy, and want to believe that escape is just that easy.

The documentary has amassed around 10 million views and I urge you to watch the film in full (Olson, 2022).

How DAOs will change the way businesses form and function

DAOs are the next biggest element of Web3 that are interesting parties for their disruptive potential. DAOs remove the need for hierarchies, which creates advantages and disadvantages. Along with no senior management, there's no middle management either. Instead, the DAO works on collective decision making across the organization using a series of predetermined rules and systems. The benefits of DAOs sound, as you might expect from something claiming transparency, decentralization, accessibility and security, almost utopian. However, despite some issues with fraud and bad actors, DAOs for the most part seem to work until greed or nefarious users come into play.

The autonomous structure of DAOs means there are no traditional management issues (conflicts, power grabs) when it comes to running the organization. Each member has equal stakes in shaping the organization, decisions and voting power. Depending on the number of tokens you have, different members may have more control over voting in the organization. In a DAO, any member can introduce new ideas for consideration in the group. The key is that all members can see who proposed what and all have a vote and opinion. Implementation of successful elements is neutrally applied thanks to the predetermined rules and systems. The smart contracts ensure accountability as every member needs to commit to evaluating new ideas and motions in order to make sure the DAO moves in the right direction. Thanks to being built on the blockchain, DAOs are documented, flexible and traceable. The latter is key for the finance market as this reduces the likelihood of successful scams and issues created by investors pulling out.

The disadvantages of DAOs are not difficult to understand. While DAOs may seem like the holy grail, they are not without issues. The most obvious is the age-old issue of power and abuse of power. Namely, if a collection of token holders band together, they can influence the direction of the company. Companies could be disrupted easily if that becomes the case. Another large drawback is the amount of time that transactions can take with a large number of people. You might not be able to implement any changes without different voting mechanics being completed – something that isn't going to be easy if you have to resolve a security issue or other vulnerability. If such an affair occurred, the organization would lose a significant amount of valuable time it could use to fix any resulting problems. The last concern also isn't a small affair: taxes. DAOs don't have rules and regulations for tax or management. Both are things that proponents for DAOs also relish. The last talking point is the dependence on smart contracts, or rather the code that underpins them. All code has vulnerabilities and smart contracts are no different. DAOs are new entities and you will have to embrace the positives with the negatives if you want to enjoy getting their full benefits. Over time, and as computing power increases, it is likely that a good portion of the negatives will be minimized.

The third area where Web3 is expanding includes Metaverse environments, albeit embryonically. Currently, there are around 40–60 million headsets in the world. While this is a large number, the majority of headsets are in the western hemisphere and the market is still early despite decades of innovation. The markets – and headsets themselves – differ greatly. From cost to comfort, access to content and entry costs, virtual reality remains a big opportunity that has a lot of issues for mass adoption.

TOP TIP

Web3 requires a mind shift and a critical eye to keep abreast of new developments. Here are several resources that offer information and perspectives to help you understand what's changing, breaking and happening next:

- Zoe Scaman writes widely on Web3, the future of fandom and lots more via Substack and on Twitter (@ZoeScaman).

- 'Zotero' has a group that is excellent for case studies and curated resources and long-form articles (https://www.zotero.org/groups/4600269/web3/library) – written by various authors.

- 'Web3.university' offers a variety of educational resources, including articles, tutorials and videos to learn the technologies that are gaining popularity in Web3 (https://www.web3.university).

- 'Web3Talks' is a podcast where Maciej Budkowski converses with founders about their projects, business models, technology, community building, user acquisition strategies, and more (https://podcasts.apple.com/us/podcast/web3-talks-stories-tips-from-the-builders/id1595653866).

- 'Web3 is Going Great', written by Molly White, a veteran Wikipedia editor, delves into the issues, dead projects, and skullduggery surrounding Web3. White does not pull her punches (https://web3isgoinggreat.com).

How will people use Web3 in the future?

The short answer is that it is too early to tell. Web3 is very much in its formative years and there are strong forces acting on the space to control it, build it and curtail it. Depending on how these forces act – and the VC money flows – Web3 will be a power to be reckoned with. Twitter and Reddit have already begun to experiment with Web3 and more is expected from other platforms and companies as elements like tokens and play-to-earn are explored and exploited.

Hundo Careers (hundo.careers), the world's first on-chain, learn-to-earn campus for Gen Z powered by a utility token, aims to end youth unemployment using Web3. At their recent CareerCon event, CEO Esther O'Callaghan OBE spoke about the future of jobs and work in Web3: 'The future will be brands and individuals creating worlds for different reasons. I have already seen dentists interested in AR/VR and environments to get people more comfortable with the aim of reducing fear and no-shows by 50%. [The metaverse] is more about giving people a valid alternative to what's there already.'

The key will be to focus on users and create new models that don't just mirror or cloak old ones. The point of Web3 is to build a fairer and less controlled internet. If the old platforms from Web2 try to control Web3, there could be lasting damage done to the trajectory of growth and desired outcomes. Currently, Web3 is reliant on a middleman of sorts – the very antithesis of what is meant to be happening. dApps rely on centralized support and services from companies such as Alchemy, Infura and Quicknode. Time will tell how fast – or how possible – it is to disconnect from the services that Web3 seeks to make obsolete.

What's the future for Web3?

The internet we have today is flawed. After innocent enough beginnings, power structures have been cemented that are difficult, if not impossible, to break. Companies have been allowed to become hugely powerful and in some cases dangerous to human life, economic stability and political democracy. Web3 offers a new way of thinking that is both idealistic and needed, but also appealing and not without issue.

Today, the web makes money via advertising, subscriptions and freemium models. Web3 offers new models and results that platforms today simply do not. A blockchain-based internet would put forward new solutions that offer creators and users methods to monetize activity and input – a stark departure from where we are today. The economy of Web3 will be tips, stakes and mining.

How will the two combine? Spotify could allow users to purchase shares in a new artist and set rules such as when the artist breaks different markets the users share a small number of streaming royalties or merchandise sold (even co-created). Meta might allow users to monetize specific aspects of their profile to earn cryptocurrency. A move like this could help Meta's efforts in creating a workable digital currency that they own, although they have been unsuccessful with early projects like 'Libra', which was later renamed 'Diem'. Mastercard has revealed plans to let its 2.9 billion users buy NFTs with their debit/credit cards – no crypto needed. The future is being written (and rewritten) right now when it comes to financing and how people transact in future worlds.

In Web3, everything can be a product or investment opportunity that is democratically governed in a way that Web2 platforms currently aren't, begging the question, just how decentralized is Web3? Currently, not hugely. The top 9 per cent of NFTs on the Ethereum blockchain account for 80 per cent of the market, valued at over $10 billion – a statistic that sounds eerily similar to how the US is currently set up, with 10 per cent owning 70 per cent of the country's wealth. If Web3 is looking to solve big problems of ownership, control and market manipulation, there's plenty to fix. Cryptocurrency could be said to continue the greatest wealth inequality anywhere in the world. Many areas of Web3 are currently being designed, developed and pushed by parties with vested interests.

If Web3 is to be equal, everyone needs to be able to access it in the same way. Due to broadband issues and lack of headsets, this is unlikely. The mobile metaverse experience is a workable halfway house, but true equality means full access and full experience. Currently, there are only 40–60 million virtual reality headsets in the world and they are not distributed equally.

Of course, this is a highly idealistic version of Web3, sketched mostly by people who have a financial stake in making it happen. The reality could be vastly different. Some founders and early movers argue that Web3 should be governed in a way that Web2 companies are not, in an effort to avoid monopolies and controlling behaviours. The majority of platforms still to this day reserve the right to snatch back usernames, ban accounts and change their rules whenever they like, without warning or recourse. As we have seen above, a Web3 approach could make this process automatic or even have users vote on individual outcomes and large decisions. Not a perfect system, but a more democratic one. The reality right now is that the web is struggling to make money unless you are the major players, Meta, Google and the like. Web3 offers a new vision where there is more privacy, fewer trackers and personal data is just that, personal.

As with any technology that promises the opposite of what is happening, it's important to retain perspective on who is pushing for the change, and the motivations behind both. Early Web3 detractor, Stephen Diehl, a software engineer in London, England, is convinced the current path of Web3 isn't all that is being promised, but the goal is solid:

> Web3 is that technical manifestation of this empty grasping for a messianic solution that's going to solve all our problems. It's entirely rational to want to build a more decentralized technology stack and to aspire to a more egalitarian internet, a more equitable society, and a better world. However web3 is not the golden path that leads us to that world, it's the same old crypto [nonsense] just packaged up in a sugar pill to make it easier to digest.

Web5 (yes, five) is already here

While Web3 isn't yet fully bedded in, the vision might be killed by Web5, an idea that former Twitter CEO and Block Founder, Jack Dorsey, is putting forward. Dorsey believes that the VC investment in Web3 makes the goals of Web3 (no controlling corporations, no gatekeepers) redundant or even impossible. Dorsey now infamously tweeted 'You don't own "web3." The VCs and their LPs do. [Web3] will never escape their incentives. It's ultimately a centralized entity with a different label. Know what you're getting into…' (Dorsey, 2021).

Dorsey is doing this via a subsidiary of Block he is calling 'TBD' (!) that offers truly decentralized… everything. Nodes, platforms, apps. Everything returns control of identity and data to the user and is built on top of Ion (an open, permission-less Layer-2 Decentralized Identifier network that runs on top of blockchain. Put another way, Web5 is Web3 without the VC-backed start-ups. Both Web3 and TBD (tbd.website) have huge issues to contend with. An apathetic global population but perhaps a more inclusive one from the start, according to the TBD website:

> We believe in a decentralized future that returns ownership and control over your finances, data, and identity. Guided by this vision, TBD is building infrastructure that enables everyone to access and participate in the global economy. TBD invites the world to join us in building and adopting open and decentralized technologies that solve real problems for real people. Our mission advances economic empowerment around the globe through the power of decentralized solutions, built open source and collaboratively. (TBD, nd)

Whether Diehl and Dorsey are right, the genie is out of the bottle. We are at a Gates' Law stage; everyone is overestimating what can be done in one year and underestimating what can be done in ten, all during a period of ginormous economic uncertainty. Conditions that create hype and overinflated expectations aren't sustainable; the best way to protect yourself is to educate yourself on the long-term (and sustainable) uses of blockchain technology. Based on what has happened and is happening already, it's clear that businesses will find the technology useful long before regular people do. If you're wearing a headset in the near to mid-future, it'll most likely be at work or you'll be gaming in VR at home.

Step one for consumers could be to create new digital worlds that don't look anything like what we have now. A reimagining of how systems work, how information is distributed and how we even exist. These worlds are called metaverses and are part of the multiverse. The next chapter explores how these worlds might disrupt, and offer opportunities for, your business, industry and consumer.

11

The metaverse – a truly disruptive technology

The term 'metaverse' harks back to Neil Stephenson's 1982 novel, *Snow Crash*. Stephenson used the term to describe the virtual place characters could go to escape an authoritarian regime. Understanding metaverses isn't overly difficult; Hollywood has been pushing the virtual world narrative for decades in films like the 'Matrix' series (1999–2021), *Surrogates* (2009), *The Thirteenth Floor* (1999), *eXistenZ* (1999) and the 'Tron' series (1982–2010), along with the newer films that mix augmented reality (*New Guy*, 2021) and full immersion in Steve Spielberg's blockbuster, *Ready Player One* (2018). The latter is where current thinking is leading, to haptic body suits and full mobility enabling people to walk around the world, fly and feel pressure when they are touched or get shot.

Rather than having one single metaverse, we have a collection of different offerings from games companies to brands and tech platforms like Meta. The term for the entire metaverse space is 'multiverse'. Multiverse is also a scientific term for a theory that infinite other worlds could exist at the same time, which has entered the public lexicon thanks to Marvel and other sci-fi films. The theory suggests we are on one Earth where humans breathe air, but there are other versions of the same Earth where we could breathe underwater, for example. This is not to be confused with 'Omniverse' which NVIDIA is pushing, which is simulation technology that connects metaverses to each other.

Definitions are still being written and will change, but these are the basic elements of a/the metaverse:

- A virtual environment that is fully immersive.

- Each human/entity is represented by a customizable avatar that represents them in some way (either in reality or fantasy – i.e. a unicorn on roller skates wearing a hoodie).

- Users experience the environments from a first-person perspective.
- Users access the metaverse using goggles, headsets and other devices.

Where are we now?

The term did not immediately take off despite Hollywood's best efforts and only saw a resurgence in usage in late 2021. According to Google Trends, the term 'metaverse' saw monthly search volumes over time near zero at the beginning of 2021 (13 December 2020–9 January 2021) and by the end (12 December 2021–8 January 2022), the term was seeing around 12 million monthly search volume over time. What caused this surge in interest can be summed up in one word: Facebook. The company announced future intentions on 28 October 2021, along with changing its name to 'Meta'.

Why is the metaverse disruptive?

A lot of elements make the multiverse and metaverse worlds disruptive, but step back a bit and think about what the creators of the metaverses out there are building before we talk specifics: places where there are multiple new alternate digital realities (and shared spaces) that allow people to socialize, work (including earning money), play and potentially lead another life. If that doesn't sound disruptive, I don't know what will convince you.

Currently, the Web 2.0 interfaces are incredibly invasive and flat, but in a full metaverse experience (i.e. with a headset of some sort), you get a richer experience that can transport you and create connections that weren't possible with websites with even the best of chatbots. Thanks to 5G and 6G connectivity speeds, the metaverse will appear second nature to people interacting within different environments, whether you're in the same town or halfway around the world. From virtual assistants to shopping malls, you will no longer be bound by physical distance, which in turn increases the number of potential connections in the world, opening up avenues for fresh thinking, new markets and new products and services.

In addition, the idea of identity will also be a huge one for people – and businesses – to contend with. From pronouns to making things attractive to different people, the sense of self in the multiverse can change from environment to environment, so companies will need to traverse this area carefully from a social and data perspective.

There will be closed and open metaverses, each offering unique issues and opportunities to disrupt elements or full businesses. Imagine a global Ford dealership that changes for every person who enters it. Ford will have that option. Equally, Ford could partner with another world, say Decentraland, to utilize their user base to create virtual vehicles to use inside Decentraland or another metaverse environment. Ford could even sell real-world versions of the vehicle if the customer wants it; he or she then gets a cut of anything sold either virtually or offline. The money and resources saved by Ford could be incredible, plus the customer connection and satisfaction levels (should all go well!) will be astronomical.

Currently, metaverses exist but are mostly in the early stages of development; the fully fledged worlds that Hollywood depicts in things like *Ready Player One* are nowhere near ready. The metaverses that are in operation (including Second Life, Decentraland, Minecraft, Roblox and Fluf World) offer varying levels of immersion (see below). The goal is for metaverse worlds to offer ephemerality or a feeling of 'live without recording'.

TOP TIP

Knowing the difference between existing platforms isn't just useful, it shows how different worlds can and need to distinguish themselves from one another. While entertainment and gaming companies are leading the way, there are others who are doing interesting things too:

- 'Fortnite' by Epic Games is mainly played on PCs and consoles and focuses on a first-person shooter-style game.
- Roblox allows players to create new landscapes to play on and users can monetize these worlds too.
- Decentraland is a completely different offering again. The 3D virtual world is owned by users and allows them to create virtual buildings and structures (think galleries, theme parks, rooms) which they can then charge entry for (powered by Ethereum).

Metaverse environments will be fraught with issues around free speech, behavioural idiosyncrasies and right-to-be-deleted elements. Speech will be the primary mode of input and interaction instead of text and typing. Another key area is representation, how we are seen and perceived in the metaverse worlds using hands, gestures and facial recognition. The syncing of these elements will be key in allowing people to connect, bond and pay attention. Currently, no system offers these elements synchronized to a degree where people are fooled or lulled into a sense that they are truly transported to a new existence or environment. There is a wall between them and what's happening. In the future, thanks to improvements in technology and sensors, the wall won't be there, and people will have new challenges to deal with. The environments offered will also allow individuals to immerse themselves to varying degrees. From the safe and understood layout of a café to the craggy rocks of a new planet, each scenario requires a different level of immersion from the user that is currently not available but will be over the coming decades.

When you factor in these elements you start to see how metaverses can and will be disruptive for multiple industries, along with offering them multiple new opportunities:

Fashion: New consumers, different price-points, new products, testing ground; companies to look out for: Ready Player Me, DressX.

Retail: New worlds and 'metaverse malls' to shop in, affiliate seller opportunities, new influencers, reduced returns; Decentraland already has a copy of Samsung's main New York City store.

Gaming: New games, new revenue streams; Roblox (+50m active users, 2022) has hosted virtual concerts already and Animal Crossing crossovers. Vans have already launched a virtual skatepark in Roblox called 'Vans World'. Also keep an eye on 'The Sandbox', an Ethereum-driven digital world.

Sports: New fans, partnerships and revenue streams; the NBA is leading the way and the Australian Open partnered with Decentraland and Tennis Australia to create a replica of the grounds for online fans.

Fitness: Lower barriers to entry, massive new worlds and opportunities to motivate people; Raramuri, a virtual sports company, held its first metaverse marathon in 2022.

Real estate: Already a booming gold rush to own space in new worlds thanks to artificial scarcity driving the prices of plots up. Decentraland has just 90,601 plots, and The Sandbox currently allows 166,464 plots of land to be bought.

Financial services: A huge opportunity for new customers and models. Beyond cryptocurrencies, how about a mortgage for that virtual plot of land or a business loan for a digital fashion line? What about insurance? IMA Financial Group has launched a lab in Decentraland called 'Web3Labs'.

Cybersecurity: A huge area for innovation (and concern). In the future, banks and providers may offer protection from ransomware attacks, theft and other issues faced by Web3 users and builders. Identity verification and privacy could also be huge areas for the industry to show real leadership. OpenZeppelin is already working on making Smart Contract even smarter.

Advertising: More, possibly infinite, space to influence others. No industry is salivating more than advertising when it comes to the metaverse. Creativity can be fully explored at lower costs and with more data. What's not to love? New opportunities like virtual influencers who can work 24 hours a day and not get tired start to come into play. We might also see a more refined approach to advertising; think areas sponsored like museum galleries with never-before-seen artworks and digital sculptures instead of virtual billboards.

Workplaces: Endless possibilities for more inspiring offices and collaboration tools. Meta's 'Horizon Rooms' are a good look at where things are now, but in the future, your office might be in a different galaxy, country or inside a sculpture.

Education: The ultimate university might just be around the corner. Imagine being inside experiments and using scale and teleportation into arenas such as the human body to increase understanding. Besides this, you could literally visit the places being discussed in class live. Educators will need to contend with safeguarding special issues around security when dealing with young people.

Events: A huge opportunity to make events more engaging, global and impactful to larger audiences. From concerts to conferences, the events industry is about to go through a revolution thanks to increased interest in VR and immersive experiences. The future of political campaigning just got more interesting too. Keep an eye on companies like Sensorium (sensoriumxr.com), Wave (wave.watch) and Mytaverse (mytaverse.com) who are trailblazing new ideas in the space.

Law: A massive new area to navigate thanks to data privacy, user interaction and private property laws all needing to be updated, and in some cases, completely reimagined. What about jurisdiction? Could a new multiverse law system be created? Digital asset theft is a real issue already; thinking about identity theft and intellectual property theft, much needs to be ironed out that will likely take a long time to make into laws, which is something brands need to be wary of.

Each of these double-digit billion, if not trillion-dollar industries are all looking for the future or the next big thing. The multiverse offers them new playgrounds and opportunities to experiment for very little money compared to real-world changes, in a safe way. While the headsets and goggles will take time to catch on (see the previous chapter for sales figures), the ability to access the metaverse via computers and other devices will bring forward new and exciting possibilities in the future.

Pros and cons of the metaverse

It almost feels too early to assign pros and cons to metaverse technology because the idea is so embryonic and going to change as people develop new ideas and technologies emerge. Still, it's important to have a plan and so knowing and understanding both sides is helpful. To that end, here are some overarching elements to think about when creating a metaverse strategy.

Pros

- New business opportunities. The biggest opportunity for most people reading this book will be the business opportunities that new areas and arenas produce. From finance to travel, healthcare to real estate, the metaverse is going to impact, and provide opportunities for, a wide range of businesses.

- No geographic barriers. You can be transported to photo-realistic parts of the world or new ones. Geography and travel have become redundant. Apart from the travel cost savings, there's an accessibility to this that will benefit billions of people. People will also be able to attend, with a click of a button, concerts and events that previously would have been impossible, which is not only fairer but also an opportunity to gain exposure to more cultures.

- Education/healthcare education innovation. From creating new classrooms that support the material being taught to personalized ones that maximize learning, the metaverse offers students new opportunities to maximize learning – the same with healthcare. Imagine having new treatment options that decrease recovery times. VR has been shown to help with post-traumatic stress disorder and a wide range of other medical issues.

- New rules, no rules. Metaverse worlds don't need to subscribe to any laws of physics or reality. You can create environments and power structures that redress balances.

- Quality of social interactions. Due to the ability to be your authentic self (that perhaps you don't fully push in the real world), people may have a richer social experience online.

- Gaming. VR headsets envelop you inside new worlds, increasing enjoyment and immersion.

- New communities. Thanks to distributed access, potentially millions more people can be brought together to help and understand one another when playing games, being educated and just socializing.

Cons

- Cybercrime. While not a new problem, cybercrime will undoubtedly impact the metaverse, as there are not currently massively sophisticated cyber security levels. For this reason, all kinds of devious antics can be popular, from fraud to money laundering to illegal services and activities. The decentralized nature of the metaverse protects nefarious types, and governments still don't have the vast resources needed to tackle cybercrime on the scale that is coming.

- Negative impact on culture and communities. One disadvantage of bringing everyone so close together, and fusing the different cultures of the world into one, is losing the beautiful cultural diversity currently present in the world.

- Identity. Linked with cybercrime, identity becomes a key currency in the multiverse. Aside from stealing people's identities, creating new ones, and multiple new ones, could lead to massive mental health issues. Aside from this, identity brokers will likely pop up to manage our identities.

- Addiction is also a concern because of the immersive and dissociative nature of the virtual worlds. People can change who they are, which, for some, is a huge opportunity to live out fantasies and become a different person.

- Loss of physical connection to the real world. It is quite possible that we won't just spend extra time with friends and family but be placed in the metaverse for days at a time. Either way, whatever time is spent in the virtual world is time away from the real world, which could break or strain a person's relationship with the physicality of the world and indeed physical health.

- Privacy and security issues. The metaverse needs new robust protocols and standards as much as new hardware and software, all of which will be hackable and open to attack from criminals.

- Mental health issues. The metaverse could give rise to mental health issues in people who prefer it to the real world when they are forced to return.

- Virtual bullying. People can be mean online, engendered by anonymity. Meta has already seen people molest others in virtual worlds, and it's just as likely bullying will transfer too, especially if digital fashion takes off and a social pecking order comes into play.

- Moderation. Most services have some form of moderation, but the metaverse will require something different. Will an Uber score for people become commonplace? Will users have to self-report people for bad behaviour? The rules are yet unwritten.

- Connectivity and hardware inequality. While we will undoubtedly visit metaverse experiences through mobile and desktop computers, the full experience will require new hardware that costs money that many billions of people cannot afford. Will this create further inequality and will connectivity issues make for a have and have-not class online?

- Corporate overreach. As with all new technologies, corporate interests and experimentation will come first, as these entities have the money and capacity to experiment and develop norms that affect how systems and ecosystems develop further.

Why Meta is (and isn't) driving the metaverse narrative

When changing its name, Meta issued a documentary-style prerecorded video featuring CEO Mark Zuckerberg, which explained their thinking about metaverses and what to expect from them. A lot looked like 'Horizon Worlds', Meta's play into virtual worlds and environments, and less about the possibilities and potential. Meta's video and viewpoint were derided by the tech world for being short-sighted and unrealistic. The videos were not actual footage, were heavily edited and led to inflated expectations, as most new technology launches do. The expectations are what experts had an issue with: in order to build and get funding, projects had to be a success and Meta's presentation added more pressure to deliver amazing experiences that the technology (currently) does not allow. Aside from that, the technology needed for the full metaverse experience is expensive, clunky and not fit for the realities of how we currently live.

Since then, after spending over $10 billion, the company has backtracked from its original announcement. A sign of reading the room, but also of changing economic and regulatory times ahead of the company. Nick Clegg, President of Global Affairs at Meta, clarified in a +8,000-word post on Medium:

> For people to actually want to use these technologies, they will need to feel safe. Companies like Meta have a lot of work to do both to build the credibility of the metaverse as an idea, and to demonstrate to people that we are committed to building it in a responsible way. That starts by explaining as best we can what our vision for these technologies is and the challenges we believe will need to be considered as it develops. It means being open and transparent about the work we're doing and the choices and trade-offs inherent in it. It means drawing on existing work to protect marginalized communities online and listening to human and civil rights, privacy, and disabilities experts as systems and processes are developed to keep people safe. And it means being clear that our intention is not to develop these technologies on our own, but to be one part of a wider technological movement.

The early spaces that Meta has provided have had to have significant safety features added and a raft of elements explained to users (a stand-out example being why avatars had no legs). Both 'Horizon World' and 'Horizon Workrooms' showed Meta's willingness to move fast and break models again. The world responded with derision and lack of understanding.

Clegg rightly points out that it is not for Meta to define what the metaverse is, create the main space or lay down any rules and regulations for them (something the Zuckerberg presentation did not make clear at all). Perhaps the step back is because people's interest is dwindling too; search traffic for 'metaverse' is in low millions every month and looming regulatory bodies are eyeballing the company. Either way, the early work the company has done will inspire and impact creativity in the space, but we are at such a nascent stage that these early issues shouldn't shape what's coming too much.

How should brands enter the metaverse?

The above is all well and good but there's a timeline and a hype bubble surrounding Web3, right? A hundred per cent correct, but that doesn't mean you don't get to sit back and relax. Instead, use the TBD framework to determine what action (if any) you need to take and when. Use the TBD compass to evaluate the level of investment you want and need to give it. Let's suppose you ran the TBD framework and it turns out the metaverse and virtual environments are something you could utilize, where should you start?

Go big (and beyond gaming)

Focusing on gaming and clubs is easy to do, the case studies are all focused on these areas, but the metaverse offers B2B an amazing opportunity to rewrite and improve how people learn new skills and are inducted into businesses (Accenture forces all employees to spend the first day on the job in their private metaverse). There's also a huge potential to re-engage users and influencers if you create a worthwhile experience to launch new products, demo products or customer service. Perhaps even self-diagnosis or resolutions of faulty products. Which parts of your business could benefit from an engaging and geographically neutral environment? After all, you're about to have access to a new world – or worlds – to observe new customer journeys, advocacy and a lot more besides.

Sponsorship is ok

You don't have to create everything at once. For some, this will be the only way their brand interacts with metaverse environments. Others will want to create their own. The difference is the cash available, desire and long-term goals.

Pick the right design team

There are many companies out there that can create beautiful images and environments, but few that do it well, and fewer still that can make sure a community enjoys what has been created. Companies like MetaVRse and Unity have digital capabilities that are forging the development of AR and VR content. Ensure you do your homework and ask future-facing questions like 'What technology are you creating that is proprietary and revolutionary?'

Decide if you can play well with others

Remember that right now there is no or very little interoperability, so if you are choosing a partner, explore them thoroughly before signing on the dotted line. Look at the number of users, demographics, dwell time and any data on the intentions and thoughts of the users. If they are open to purchasing, that may be more important than a larger number of users. Not every company can do what Nike did and buy a start-up that creates virtual trainers, but for Nike, it made perfect sense. Could something like that work for your company?

Plan, plan, and re-plan how you turn up

New spaces are hard, norms need to form; there are right ways of doing things and there are very wrong ways of doing things. If you don't have the right back-end support to make a new community work, pause and rethink launching. Also, launching with brand advocate support is a good idea. Think about messaging, how much transparency there will be and what issue resolution looks like before launching. Think about reimagining hospitality from the off.

The metaverse is being created before your eyes. No one has an end-point or a roadmap, so create something that will a) last, b) be flexible and c) can be built on. The world is your oyster, but check that the world wants that oyster to begin with. For many brands, it is test-and-learn time. For others, observing from the sidelines is the best advice for right now while edges and norms form. While we are years away from an iPhone moment, the day Apple gets involved with goggles, glasses or a headset is when the metaverse era will really begin to fly. Before then, an era of augmented reality from companies such as Snap, Meta, Amazon and others will push new boundaries and explore new opportunities that the metaverse will utilize when the time is right.

CASE STUDY
'Shady Doggo' by Balenciaga

What

Like many high-end fashion brands, Balenciaga ventured into the metaverse early
and was one of the first, if not the first, to fully utilize Fortnite. Players could
purchase four digital outfits inspired by real-world Balenciaga pieces from a custom
virtual boutique. The digital collection augmented the wardrobes of popular
Fortnite characters: Knight, Ramirez, Doggo and Banshee.

How

Various items, Balenciaga's Triple S Sneakers, for example, could be unlocked
through gameplay, but the majority had to be bought. Balenciaga added an element
of scarcity as the hub in Fortnite was only open for seven days. Players got to try
on 'Shady Doggo' outfits and hang out, but the focus was squarely on purchase.

Results

Aside from the PR surrounding the launch and the week surrounding the activity,
Balenciaga sold thousands of dollars of clothing – both virtual and physical (there
was a limited-edition run of Balenciaga x Fortnite hats). Some items could only be
unlocked if a physical purchase was made. Balenciaga also maximized the runway of
the activity by creating an offline element in the real world on massive 3D digital
billboards in London, New York, Tokyo and Seoul and subsequently promoting videos
of the billboards online and through social media. As a result of the success of this
campaign, and others since, Balenciaga is now working on its own metaverse
'development' and has a dedicated team, according to CEO François-Henri Pinault.

Fashion has embraced Web3 and specifically metaverse environments thanks
to the avatar element most, if not all, require. Avatars can be styled simply
by default settings but increasingly platforms are seeing not just new revenue streams but new forms of self-expression emerge. From a virtual 3D
catwalk with a physical runway for Dolce & Gabanna's 2022 collection at
Milan Fashion Week to virtual-only lines, fashion is already pushing boundaries and getting data on what people will (and won't) pay for. 'The usability

[of digital fashion] is the point that's missing, but that's making gigantic steps every day,' Balenciaga chief executive Cédric Charbit said during The Business of Fashion's annual Voices gathering in 2021.

What comes after the multiverse?

Web4 – or where brain–computer interfaces (BCIs) come into play. Think of this as the wild world of wet work (wires and chips in your brain) and technology combining. In this world, you'll have ports in your brain and digital contact lenses and other technology that interacts directly with your brain. BCIs will allow commands to be executed quicker and allow humans to do more, faster and easier. At least, that's the sell. We are in the extremely early days of this actually being useful, although Elon Musk is already working on this at Neuralink and has had some controversial success and a lot of dead animals. Neuralink has claimed monkeys can play the computer game Pong using just their minds and parrots can move a cursor onscreen without using an input device, just their brain. In this Web4 world, computers talk to each other and make decisions for us.

This chapter looked at the future of the internet, or rather, the next evolution if a lot of variables and ideas come to fruition. The roadmap is not final, routes and the landscape will change, but that's the beauty of Web3; there is much currently wrong and a plethora of work to get right. Knowing what to look out for is key, and the next chapter focuses on additional elements your company will need in order to make it to 2025 and beyond with the least amount of disruption possible.

12

Disruption and newer generations

Millennials are about to be the biggest group in the workforce and this poses a unique opportunity (and threat) for employers and businesses. This generation is rewriting the book on employment, retention, training and shifting purchasing power. Never before has there been a more disruptive generation, thanks to the tools, abilities and power that they have at their disposal. Disruptive technologies are what this generation grew up on, and how you use them will be a big part of your success.

Specifically, in the next chapters you will learn:

- how to harness the power Millennials have for your business;
- the future of the Millennial workforce.

One of the – if not *the* – largest generations in history is about to move into its prime spending years. More than just the simple consumers of previous generations, Millennials are different. Millennials are already beginning to reshape economies; their unique experiences will continue to change the ways people buy and sell and companies will need to continually adapt. This is the generation that might just say 'enough is enough, here's where we need to be and here's how we get there'. Don't misunderstand; like any group of people there will be different groups and elements that jar with this image but never before in history have we needed such a perfect storm of technology, behaviour and data to enable change on a massive scale if the systems already in place allow it (or perhaps even if they do not).

An important question for businesses to ask now is: 'Who picks up the baton next?'

Generations have always fascinated, annoyed and plagued me. Roughly speaking, the previous generation 'names' the next and someone somewhere along the way gets rich. Generational titles have been around for a number of years and, speaking with several experts, no one really knows why they came about. Listed below are the most commonly used generational titles today – more are undoubtedly on the way as the world and pace of change increases (Sanburn, 2015):

GI Generation: 1901–1924.

Silent Generation: 1925–1946.

Baby Boom Generation: 1946–1964.

Generation X: 1965–1979.

Millennial Generation: 1980–2000.

Generation Z: 2000–2020? (No one has offered anything different... yet!)

Generation Alpha: 2020–2040.

This chapter will focus mainly on Millennials, but to some degree think of the word Millennial as a reference to younger or newer generational groupings. The term 'Millennial' echoes down the corridors, decks and conference rooms of agencies and brands across the world – and with good reason, as we'll see shortly. It is important to note, before we get into detail about this demographic, how the term came about. Neil Howe and William Strauss are famed for coining the term 'Millennial' in or around 1999 and therefore got to decide the traits associated with the group. If you were born between the years 1980 and 2000 you are dubbed a 'Millennial'.

A quick Google search and flick through most media outlets reveals a great deal of agreement when it comes to defining this generation. Summarizing these findings, Millennials are a generational subset of the population, made up of individuals who are:

- lazy;
- obsessed with technology;
- living with their parents;
- narcissistic;
- overly ambitious.

The trouble is... Millennials don't really exist.

Why saying 'Millennial' is not helping you or your business

Beyond making a speaker look old, saying the word 'Millennial' or any other generational title will, more often than not, make them look foolish.

Generations simply don't exist – they are made-up names for stereotypes and the grouping together of people who are often vastly different, within the same country, let alone in other countries around the world. Although it is highly unlikely that there is not a single person born between 1980 and 2000 who doesn't exhibit some of the traits listed above, the same could be true of people born 1950 to 1970 (or any other period in time). There are many other points to consider when using generation generalizations. Most – if not all – census bureaus do not define generations. Beyond this, using generalizations isn't helpful to the question often being asked or the problem looking to be solved. Often the opposite occurs and we simply tend only to reveal our own prejudices and opinions about the people we are attempting to describe.

TOP TIP

Take five minutes now to think of five better terms to use when describing age demographics or target groups. Here are two to get you started:

- People born before/after...
- Your parents' parents.
- _____
- _____
- _____

History tells us that all new or emerging 'generations' are usually somewhat negatively perceived by the incumbent generation – especially regarding their attitudes towards technology. Many marketers struggle with new demographics in general when they emerge but Millennials are particularly perplexing because of the transformative nature of the technologies and ideologies they have given way to.

The problem with these classifications is that they don't really help anyone and often lump people together to normalize behaviours, for speed rather than accuracy. Speaking at 'Deep Shift', a marketing conference, Adam Conover (the US TV presenter) summed this up: '"Generations" are usually just old people talking smack about young people.' This practice has been going on since older people have had younger people to talk about. This idea is key to understanding and challenging our own biases, predetermined attitudes and schemas (learnt responses) to get to the truth or correct decisions.

Challenge your myths about Millennials

In order to fully understand disruptive technologies you must challenge your beliefs about them and who uses them. Understanding Millennials will be a huge part of your business's future strategy and success whether you like it or not. The B part of TBD is incredibly important here because the new wave of young people coming into the workforce have vastly different ideas and ideals based on the technology they have at their fingertips, have grown up with and in some cases have even helped create. Charlotte Burgess, a Senior Director for C Space (a global insight, brand and innovation consultancy) was asked to focus on this group:

> Where former generations were happy to get a sense of meaning and purpose from the church or state, the Millennial workforce today expect this from their work. So they seek to join businesses which have a bigger societal impact (or that give them some sense of a bigger purpose), and this gives them existential meaning. (Burgess, 2016)

It is important to focus on this group because of the unprecedented changes to the way information is shared and goods are bought (among other things) and the other technologies that will be enabled by existing technologies in the coming years. Let's look at debunking some of the big myths surrounding this group before we look at why they are so important to the future.

Millennials are no lazier than previous generations

From *Time* magazine covers ('The Me Me Me Generation') to major socio-demographic reports from Pew and others, Millennials are dubbed as lazy and entitled but these images are not representative of the group. Regardless of whether you believe this to be true or not, the biggest point of interest for business owners is that the group is about to become the largest demographic in the workforce. Your workforce is about to change dramatically – how you handle this change will radically impact the course of your business.

Millennials display narcissistic traits more openly, but they are not necessarily more narcissistic

The 'narcissistic' label is hard to refute in this world of selfies, status updates and tweets. A 2010 study by Trzesniewsk and Donnellan looked at this element but found no significant difference: 'We find little reason to conclude that the average member of "Generation Me" is dramatically different [for narcissism] from previous generations.' In other words, behaviours manifest beliefs in other ways. While younger people have more technology than previous generations to utilize and create things with (i.e. followings, content, connections), the same life stages occur and attitudes change accordingly.

Millennials love technology... just like every other generation before them

It is important to remember, and identify, that it is technology changing society and the world, not Millennials – that would be too much of a stretch, for now at least. Whether we look at the video camera, the computer or the pocket calculator, every generation has had 'their' technology; this group has had more than most as the pace of technology has advanced significantly but there is undoubtedly more to come based on the technologies we are seeing unleashed today and in the near future. Author Douglas Adams (2002) summed up different attitudes to technologies:

> Anything that is in the world when you're born is normal and ordinary and is just a natural part of the way the world works. Anything that's invented between when you're fifteen and thirty-five is new and exciting and revolutionary and you can probably get a career in it. Anything invented after you're thirty-five is against the natural order of things.

Millennials live with their parents

A significant amount of people in this group do live with their parents but this is through necessity rather than choice, due to a variety of factors but mainly economic ones. Like most people born before them, 'flying the nest' remains a rite of passage but due to the choices made by previous, mainly living generations, those born between 1980 and 2000 are the most economically challenged in recent history. This should concern every business owner and brand out there as this fact will have massive impacts on consumption trends and attitudes.

Millennials are overly ambitious

I hear this a lot when talking to managers and CEOs of brands I work with and it is an interesting observation. In some respects, I agree – many of this group are overly ambitious for the level or role they are in but when you look at the circumstances they find themselves in, would you be any different? They face massive financial uncertainty, have a strong desire for a work/life balance, and have multiple working scenario options. Multiple studies suggest this drive for early success implies their primary motivation is strong family commitment and a desire to provide for them in the long term.

Ambition in and of itself is neither a bad nor a negative thing but if left unchecked this behaviour can become problematic. However, is this any different from previous generations? No. Every generation has been labelled 'entitled' and 'ambitious' since the dawn of the terms themselves. Would you prefer a workforce that wasn't ambitious? Do you think your company would thrive or be in the same position in five years' time if the workforce you have was less ambitious? Be careful of making this mistake. Be open to ambition and cultivate it – this may mean changing policies or creating new ones.

Based on these points you could conclude that generational stereotypes are about as useful as horoscopes. Whether you agree or disagree with me is not particularly important but the way you approach individuals born between 1980 and 2000 will, for a wide number of reasons, determine the success of projects, initiatives and ultimately your business. Dan Keldsen (2016), author of *The Gen Z Effect: Six forces shaping the future of business,* summed this up to me nicely:

> The number one myth about Millennials is that they are any better or worse than any other generation, young or old. The entire philosophy behind the 'Generational Gap' (popularized in the 1960s by Margaret Mead) was an off-hand remark; it was never meant to be etched in our collective minds that 'no two generations can ever stand each other, let alone understand each other.'

Why are people who were born between 1980 and 2000 so important?

The one thing you can say – without doubt – about the group of people born between 1980 and 2000 is that they are the most diverse group of people born in a specific period for at least 100 years (certainly for most Western cultures):

> Young people are more tolerant of races and groups than older generations (47 per cent vs 19 per cent), with 45 per cent agreeing with preferential treatment to improve the position of minorities. This may be attributable to

the diversity of the generation itself, which recalls that of the silent generation. The shifting population is evidenced with 60 per cent of 18–29-year-olds classified as non-Hispanic white, versus 70 per cent for those 30 and older. This reflects a record low of whites, with 19 per cent Hispanic, 14 per cent black, 4 per cent Asian, and 3 per cent of mixed race or other. Additionally, 11 per cent of Millennials are born to at least one immigrant parent. (US Chamber of Commerce Foundation, 2012)

Racial diversity is a huge element of disruptive technologies – different backgrounds, different upbringings and different interactions will lead to new ideas and strategies forming.

This diversity in and of itself has enormous power to change many technological, cultural and international issues that are present in society today. However, it is important to note that this power has yet to be fully realized and wielded by anyone, let alone young people. Any business owner should realize that this will change drastically. Policies and practices will need to reflect this diversity if a company is to look appealing to the new wave of candidates, the next wave and the outside world.

In 2015, a tipping point occurred in the United States that had already happened in other parts of the world and is taking place in others as you read this: Millennials now outnumber the previous generation in the US workforce. More people born between 1980 and 2000 entered the workforce than from the previous generation, making them the most powerful force in the labour market. Diversity and workforce aside, this group of the world's population has several other interesting points – political power being one that is fast going to become an issue.

The demographic shift has huge implications for the business world beyond simple numbers and human resource issues – this group of people thinks differently because of the world they grew up in. Previous generations were defined by physical wars; this generation has technological ones. It is the speed of technological change that is really empowering this generation and increasing the potential of what they will achieve. Disruption is the normal way of life for this group and as such, static businesses and structures seem antiquated and are rebelled against or avoided in favour of new ways. Being the most educated generation to date, those born between 1980 and 2000 are in more debt than previous generations but are also the most forward-looking, with over 70 per cent of them already saving in some way for retirement.

Adam Smiley Poswolsky, author of *The Quarter-Life Breakthrough: Invent your own path, find meaningful work, and build a life that matters*, explained to me why Millennials matter:

> The average American is staying in their job just five years. The average Millennial is staying in their job about two to three years. Already, 34 per cent of the American workforce is freelance, and this number is only expected to grow. The traditional corporate career ladder is dead. For the average millennial, a job is a brief experiment and learning experience. (Poswolsky, 2016)

Poswolsky went on to describe how businesses should approach Millennials:

> The companies that want to attract the top millennial talent will have to engage and train employees for a short, two- to five-year learning journey (and then help them find their next great opportunity). They will have to understand and accept that no one wants to stay at one company forever. Employers can think of their employees as not just independent contractors, but as dynamic individuals with long-term career and personal ambitions. If they help someone move on to something even better, that former employee will continue to have brand loyalty and may even recommend their friend for the job they just left. It's going to be a much more fluid system of sharing and investing in talent.

The comments above should concern any business owner; this is a huge shift in thinking, resources and the status quo. These concerns highlight how important it is to attract the right thinkers and talent to your business in order to maximize the potential for disruptive technologies but also not be blindsided by them. How your business prepares itself for this shift could mean the difference between being in business and not. When it comes to 'Millennials' the name may be irrelevant but the behaviour and the shifts this group will continue to cause are anything but.

Beyond these changes, however, is the future for this group – a drastically different one not because of wars or economic crises (both have been norms for this group). When you look at the leaps and bounds science and technology are allowing, the future for this group could be truly fantastic – from cryptocurrency that could revolutionize everything from banking to buying a coffee, to Tesla's great solar power plans. These technologies alone have the potential to impact business through increased customer spending thanks to more trustworthy online purchases and more expendable income because of lower bills. The next two decades are extremely likely to see a large share of previously impossible leaps – many in part made by those born during this time. According to Benjamin F Jones in his paper 'Age and Great Invention' (2005), the peak innovation age, based on Nobel Prize-winners, is between 30 and 40. The argument here is that while age brings wisdom (or at least as the old saying goes), the opposite may be true for innovation and openness to disruption. Data in studies conducted around CEO age and innovative behaviour has shown that younger CEOs file more patents than older ones and also hire younger innovators – a trend that has massive ramifications for the older workforce. If your company can't attract young talent you are in trouble, not just immediately but further down the line too.

Your business needs EQ not IQ – form a Millennial mindset

Millennials are the first generation to grow up with the internet as standard. Let that hit you for a second. Open source, collaboration, unlimited knowledge and access has been a 'right' for this group since they were old enough to know what a keyboard is. Because of this, Millennials are fast adopters (and deserters) of technology that helps them achieve their goals. Instead of simply accepting the status quo, this group are dealing with the sins of their forefathers and saying there is (or has to be) a better way. This is Emotional Quotient (EQ) – a person's ability to work with others, understand them and what motivates them. IQ remains important to them but EQ is what sets them apart. This EQ is partly due to their diversity but more because of the issues they face. EQ is an important factor when thinking about TBD and disruptive technologies because of the behaviour element of TBD; understanding what people need is an important skill when it comes to spotting or creating opportunities to use disruptive technologies.

The ramifications for businesses are clear:

- new training will be required to make sure people understand each other;
- different managerial styles and structures will be needed to make sure people can work together effectively;
- greater transparency will be needed in order to attract and retain this group – think about how open you can be and how you will demonstrate this back to the group;
- opportunities for collaboration, job share and upskilling will be required to interest this group;

- new HR practices will need to be created and developed – this group like to recruit for you as much as they will leave you when they have learnt everything from the opportunity;

- open systems and mentalities will need to be developed to handle the innovation and new ways of doing things this generation will bring to your company if you let them.

Letting them is of course the key here – working with both large and small clients with issues in and around this area has opened my eyes to the challenges businesses face with this generation and generations to come. Do not dismiss the disruption this group can, and will, cause your business from the inside or the outside. By enabling individuals and having a happier workforce, businesses will be able to focus on bigger issues, take bolder steps and make strategic headway in uncertain times using the right technology and making the right technology bets and decisions.

Equally, think about Gen Z in the workplace as a different breed from Millennials. Interviewed on 'Mouthwash' (TBD's audio show), generational expert Dr Eliza Filby shows how Gen Z's experience during the pandemic (their formative work years for many) differed from Millennials.

> Most Gen Z-ers' work experience comes from the pandemic. [Gen Z] see the curation of their social media identities as something that is about controlling the image. Zoom, on the other hand, is less comfortable and arduous... a less airbrushed experience for them. These 'digital natives' prize more than any other generation face-to-face contact, many don't have home offices, many returned home and many found their mental health suffered during the pandemic. Millennials, on the other hand, had a very different experience because of their position in the workplace, expectations and economic realities. The next 15–20 years will be very different when it comes to the workplace because we have a generation that has grown up thinking in a much more agile way, whether that's education, money or their career. We have a generation that has been groomed to think in a much more agile way about their career; a much more dynamic, multi-stage career.

TOP TIP

Consider eliminating the annual review – instead perform quarterly check-ins. According to American Express, only 48 per cent of companies actually complete annual performance checks and these increasingly frustrate younger members of the workforce as a year is seen as an eternity with the speed things are happening around them. By setting smaller, achievable goals, not only do you increase year-round motivation, you also allow all groups – not just Millennials – to course-correct and proactively deal with challenges as they arise rather than wait and have issues or frustrations fester.

Do not dismiss the desire for meaning and meaningful work

As mentioned when we discussed the myths about Millennials, working for a meaningful goal or employer is one of the largest motivators for this group. Peter Diamandis, Chairman and CEO of the XPRIZE Foundation, calls this a massively transformative purpose or MTP. A good way to think about MTP is as mission-driven but not a mission statement. Instead, most MTPs are guiding principles rather than simply good straplines.

Some good examples:

TED – ideas worth spreading.

Quirky – make invention accessible.

Tesla – accelerate the world's transition to sustainable energy.

Google – organize the world's information.

Airbnb – become a trusted community marketplace for people to list, discover and book unique accommodations around the world.

Cisco – connect everyone and everything, everywhere, always.

Singularity University – positively impact one billion people.

Unique. Lofty. Bold. Believable. These MTPs have lots of things in common and it is easy for big companies to say such things but as Diamandis points out, this focus, goal or big-picture belief symbol is what drives a lot of the people coming into the workforce. In their book, *Bold: How to go big, create wealth and impact the world*, Diamandis and Kotler explain how to determine which is the right MTP for your company and why it is so important to make sure the company's purpose is understood by all:

> Autonomy is the desire to steer our own ship. Mastery is the desire to steer it well. And purpose is the need for the journey to mean something. You're not going to push ahead when it's someone else's mission. It needs to be yours.

Retaining younger demographic workers is, and will remain, a core area of interest for business owners for years to come as business trends change, people evolve and new technologies alter career paths and opportunities. This does not mean it is impossible. Here are some easy wins I have observed through my experiences working in-house, at agencies and with big and small brands – ways to retain young employees and utilize the resource rather than being swept away by them:

- Create the *right* environment, not the coolest. Silicon Valley and the media have perpetuated the stereotype that an office needs to have skate ramps, foosball tables and a barista. These are nice to have but obviously not everyone can, does or want to work in a warehouse. Millennials are looking for employers that embrace the freedom to move about rather than being sat behind a desk.

- Challenge them. Use the traits of the group to your advantage by giving them the authority to challenge you. This may seem counterproductive but challenging them, or a small group, to prove why something is better done another way can be time better spent than employing 'experts' who will agree with you.

- Create a free innovation team. This group want to impress and are technologically advanced. Allow them the freedom and ability to test new technologies and make suggestions to you about what to implement. Whether it is augmented reality, 3D printing or machine learning, the chances are they know more about it than you and may even be able to introduce you to an expert via their extensive 'life' networks.

- Flexible hours aren't a joke; they are a way of life. The elusive 9 am to 5 pm seems to be a running joke in several agencies these days, which is dangerous. Instead, a better way of thinking about time is measuring it in tasks that are completed rather than hours that are worked. Some of the best young minds send me emails at 1 am because that's the way they work or when something came to them. Give this group the freedom to work in whatever way they choose (within reason) and get out of the way – it might be the best decision you ever made.

- Fight the one-and-done training. Employee development programmes should be an always-on affair. Whether it is shadowing, job share or co-working, you should be changing working roles up and making sure this group feel challenged and inspired to look forward.

- Motivate in the right way... which includes money. Millennials can be as or more principled and mission-led than any generation that came before them but they still have bills to pay. Instead of simply rewarding with a salary, though, think about what else you offer them. From prizes to friendly competition, some of the old tricks, done in new ways, work extremely well.

Conclusion

Millennials (and Gen Z and Alphas) will remain a subject of derision as long as people allow it. Always remember that writers get rich through naming generations – fight against using and perpetuating the stereotypes. Unleashed, this and subsequent generations have the potential, drive and ambition to make big changes if we get out of their way. We live in volatile times where one tweet can take down an entire brand or a single video can knock 62 cents off every dollar of a share price. The last reference is an example I cite regularly because it was the perfect storm of TBD. Domino's Pizza suffered embarrassment and a massive 10 per cent stock drop when a film of two employees doing unsavoury things to a customer's pizza went viral on YouTube.

Younger generations expect a lot and why shouldn't they? They're ready to give a lot in return and know what is on offer in uncertain times. I ask executives who talk about younger generations as a huge problem, 'Would you act in a different way if you had grown up as they did?' As you travel through the uncertain, choppy waters in the coming years, having the right team with you makes sense – a steady hand, a map and the people who challenge you to get to your destination more efficiently. If you have an open mind, you might find younger generations won't just treat you as a place to learn something; they may help you fight to keep the lights on and want to stay for longer.

13

The future of TBD
and disruptive technologies

In this final chapter, we will draw everything together and look forward. Specifically, after reading, you will be able to understand:

- why success in the future means flexibility now;

- why TBD will travel with you;

- what future elements may be added to TBD;

- why the future shouldn't be feared.

Before we move on to the future, it is important to look at where we are now so we know where we want to go and what it will take to get there.

Aside from coming out of a global pandemic, we are living through a massive period of uncertainty, including:

- tectonic changes to social and political ecosystems;

- a media ecosystem in turmoil and disarray;

- wars in several nations;

- multiple digital platforms bigger than most countries that are controlling information in various ways;

- climate change on a scale previously unheard of.

This combination would usually be enough to make any business entrench and take stock but increasingly – due in large part to disruptive technologies – this position is untenable. Technology is progressing much too fast and businesses risk being left behind and dislocated by incumbents in the market, let alone emerging competition. The world is not as stable as it was 12 months ago, and this situation is unlikely to change significantly in the following 24, possibly more, because of the far-reaching impacts of recent events. Disruption from technology, political change and economic change is nothing new but the effects can be faster and wider reaching thanks to new and emerging technologies. Technology has had a large part to play in recent events, whether economic, political or simply spreading information; the next few years will be critical for most businesses as they look to grow and move forward in uncharted waters.

The future could be said to be highly unstable and undesirable but before you close the book and begin rocking, pause for a moment. We live in a time where space tourism is a realistic possibility for the first time ever, renewable energies are finally becoming cheaper than fossil fuels, smartphone penetration means continents are more closely connected than ever and the poverty gap is decreasing. All of these issues require us to be one thing – flexible. The reason for this is simple: this is what's coming, and what's coming after that is going to be exciting, scary and exponential.

With such uncertainty looming, the best advice is usually to make decisions that enable you to be more agile and able to absorb knocks that will come at you. TBD is a flexible framework that can be used as is or altered to fit the needs of your company today and tomorrow – whatever comes your way. Disruption (and the increasingly used 'dislocation') seems to be coming from all sides and for many it is. For many, however, it just feels like it is and it is important to know the difference. The framework and the two versions of TBD enable users to have a more controlled and focused approach to the future. This control and the ability to remain agile are what will set your company apart from other businesses.

The future is rarely certain, but with recent events, this statement has never been truer than it is today. Seismic shifts have happened and will continue to happen thanks to the fallout of these events, and how you handle them and consider them will determine your success. The TBD framework was developed for ultimate flexibility for such scenarios. Both versions of TBD were designed so that such events can be factored in and adjusted in order to make strategic decisions faster and retain agility.

TBD is a framework – you are what makes it work

The decision to fly to the West Coast of the United States to start my career wasn't a well-thought-out one but it was a defining one. Clients, friends and family tell me a lot of the drive and energy I have comes from my time in Los Angeles but my love for technology lies at the feet of my father and his father. Whether it was upgrading the computer, a new car in the driveway when we went to see the grandparents or going to a Curry's to browse, somehow I'd find a way to evaluate it. While I am sure both had an impact on the way I view disruption – and indeed how I go about solving problems for myself and clients – I put it down to two simple things: limited time and choices.

Chapter 2 discussed how to manage your time better but many people don't know what they are managing it for. I urge you to think about the career you are in carefully because of the reason that keeps me going. Limited time.

Let's put this into perspective. Take a piece of string or rope and look at it. It is a continuous line that has a start and an end. This string represents your life. Now take a pair of scissors and get ready to make some cuts.

Cut off a third of the string – this is the amount of time an average person spends sleeping. Now cut off a quarter of the remainder; that's your child-hood and early education. What you have left is the time you have at work and retirement, except it doesn't factor in exercise and entertainment. Cut off another third of what is left; this third represents the time an average person spends at the cinema, all the TV time and the 'extracurricular activi-ties' people get up to. You should be around the 21 years left mark. That's not bad, right? Twenty-one years doing what you want to do? Wait, don't forget eating and drinking (eight years), waiting in lines (five years) and things like housework (six years). Added up, these equate to about half of what's left. Chop off another half. The rest is what's left for you to work. All the string at your feet is taken; what's in your hands is what you decide to do with it. You have roughly 10.3 years to make a difference, make a mark, take a stand, change things, leave a better place than you were given. But will you?

Let's say you live to 80 – some argue more, some argue less but an aver-age in the UK is around 80 years. If this was represented on a page it would look like Figure 13.1 – your age on one axis and each week of your life on the other axis.

Find out where you are now and cross through the weeks that have been used. Now cross off the final 10 years – people aren't usually that productive in this part. Now think about all the ways you use your time and put lines through it – it sure goes quickly. Some of it is already gone, other parts of it won't be productive or... gulp... you may not get to use all of it. The shaded bit in the middle, that's now for most people reading this book, is what matters, or is at least when you can have the most impact doing what you love to do.

FIGURE 13.1 Your lifetime in weeks over 80 years

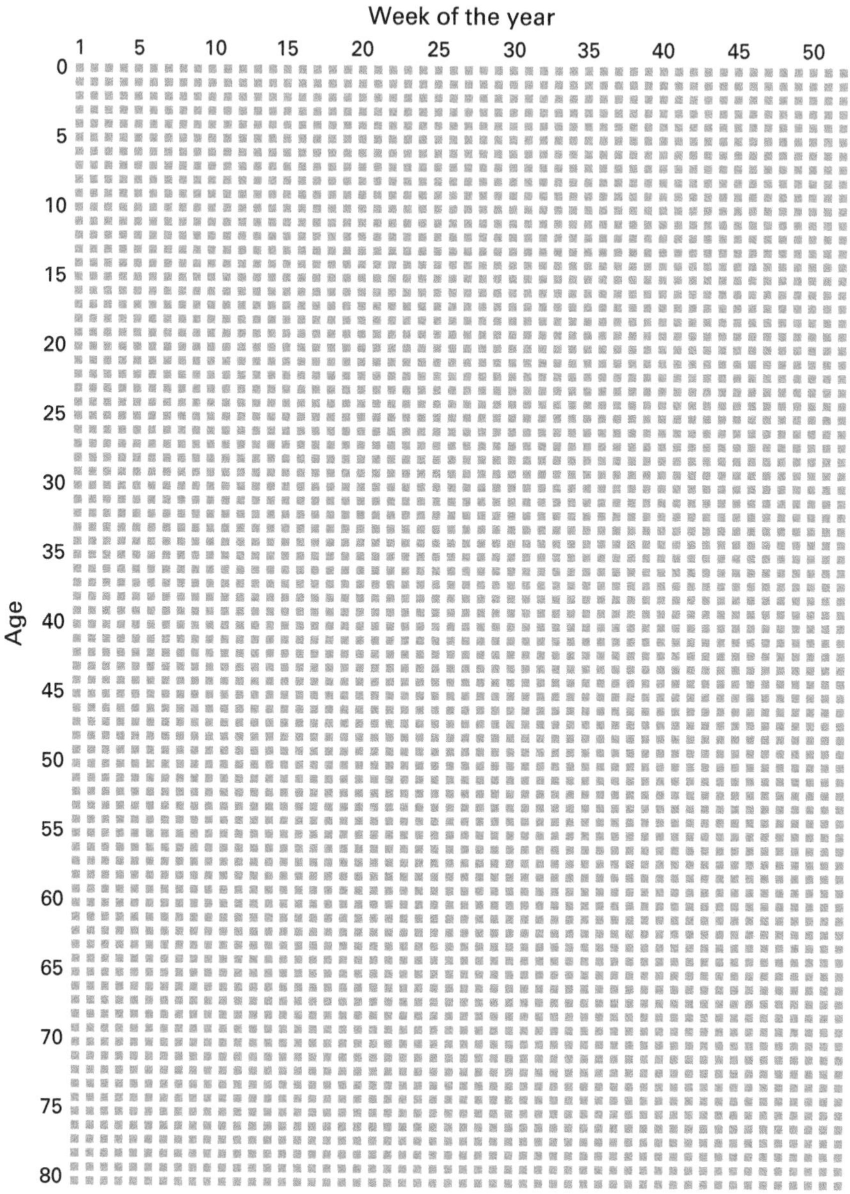

How much time do you have left?

How many weeks do you have to make changes?

How are you going to make everything count?

The point of this is not to depress you but rather to focus you on spending the time you have on this earth wisely. Regardless of whether you believe in a higher being or not, everyone gets a set amount of time on this rock, and what you choose to spend your time focusing on is totally up to you.

Why TBD goes with you through life

The earlier chapters are key to really understanding disruptive technologies and creating the time and attitude to do something about them. Knowing that your time on earth is short is one thing but understanding actually how short the potential is for you to change things and impact lives is quite another. Individually, time spent at each job is shortening – especially for younger demographics. Understanding that the time you have to make changes and implement ideas is always decreasing shouldn't depress you; it should motivate you to make bold choices and smart decisions. Time is a huge part of what disruptive technologies not only require but demand because of the times we live in and the changing variables that act upon us. Knowing all this should spur you on to focus your time on achieving and trying great things. I know it certainly does for me and the clients I get to work with. Make it count.

Clients tell me they often find themselves using the simpler version of TBD to make daily choices. One even went so far as to create a writing pad with it on instead of writing out three columns and evaluating ideas and platforms before decisions are made so it became habit. Beyond this, though, simple TBD does enable you to do one simple thing: commit to action. Clients tell me that motion, or even just the appearance of motion, is often the biggest element to success or big leaps forward, so making things visible and easy to understand and communicate back is critical. TBD helped them do this because it was simple and flexed as the situation required.

TBD+ is still a flexible framework but the focus is different. Due to the way it is set up, TBD+ brings companies closer together but also produces robust results. Results from clients are varied – it works really well for some while others simply find it a great way of coming together to forge new paths or correct old ones. Beyond this, if used correctly, it can continually offer new ways of thinking and doing things. People tell me it is because of the flexibility and continual refreshing and progress-checking process that TBD moves with people as they progress (upwards and onwards).

Beyond the TBD frameworks, success relies on the commitment to do things differently, work differently and focus differently. Understanding this is also a huge part of making disruptive technologies like blockchain and 3D printing work for you. The status quo, 'because we've always done it that way' and the non-boat rockers will increasingly find it hard to fit into the world that is emerging. You don't have to be first movers to gain value from the technologies that are coming over the horizon.

The future should not be feared

In times of political and economic uncertainty one thing is to be expected – fear. Fear of the unknown is nothing new but it is important to recognize it and deal with it. Fear manifests in lots of ways; anger, pauses, denial to name but three. The issue with fear is not to avoid it but to work through it and mitigate issues before or as they arise. TBD is a good, flexible framework that bends and moves with a company, allowing for any eventuality to be explored in a safe and positive way. Use TBD as your guiding light and make disruption work for you. We live in exponential times and this should be celebrated. In previous chapters, we have looked at ways of predicting and avoiding potential roadblocks, but knowing you will never predict or avoid them all should give you a sense of calm – safe in the knowledge you can at least predict that.

Fear is a natural emotion. Fear is a survival instinct and one of the easiest emotions to trigger. Recent events have a lot of fear circulating, some perpetuated by powerful technologies (internet, mobile) that did not exist 30 years ago, and this only adds to the confusion, unease and concern. The potential for these technologies is great but it would be poor judgement if we did not consider all sides of the argument – a core feature of the TBD framework and methodology.

Take artificial intelligence, for example; we simply do not know how this technology will 'evolve' but most agree we are at the early stage of a massive period of change – some good and some that will see jobs change. I discussed the future of AI with Jason Pontin, Partner at DCVD and previously Editor-in-Chief and Publisher of *MIT Technology Review*, and he agreed that the most important technology for digital-oriented business will be the domain of artificial intelligence and 'deep learning'.

Deep learning is a term applied to software that attempts to mimic what happens in the neocortex of our brains – specifically the activity between the layers of neurons. 'Learning' happens when patterns are recognized from digital images and sounds, and can be correctly identified by the computer. Pontin believes:

> We are already seeing the effects of deep learning in several areas – not all
> of them positive. It's very hard to guess the full impact of deep learning; the
> technique has already made unprecedented advances in image and speech
> recognition, translation, and predictive modelling. It could be that 'unsupervised
> learning' will allow really startling breakthroughs. These advances in AI
> and potential breakthroughs from unsupervised learning will suggest many
> innovations to companies while continuing to place downward pressure on
> the wages of middle-class workers around the world, because they allow the
> automation of many jobs.

Blockchain is another example. There will be many implications and manifestations before we consider this technology 'mainstream' or indeed are close to seeing the 'full impact'. Job loss is a core concern for many individuals out there who are usually of a certain age and skillset. Understandably, fear mentality may present itself and it is important never to forget the human element of any and all change – hence the importance of the B in TBD.

Uber has a chequered past and an interesting future but the beauty is not the company itself but the ecosystem it creates and enables. Before the pandemic, Uber was just dabbling in food delivery. Now, the company has a fleet of robots to deliver food, numerous ghost kitchens and robust deals with supermarkets on multiple continents. Amazon too is now making large plays in the grocery industry. The area, thanks to the pandemic and new technologies creating new behaviours, is being disrupted and will continue to be so for at least the next decade. However, regardless of it dominating the headlines, in the long run, revenue and profits will always matter more in the current economic systems we are embroiled in. While Uber is not a great example of disruption per se, it is a great example of a company that could soon be disrupted. Thanks to legislation challenges and public opinion, several governments are now looking very closely at the company in both a positive and a negative light. Beyond regulation, of course, we have other issues to contend with – the human element can never be discounted.

The issues that matter are changing; platforms and companies were being built, whereas now we are seeing what can be built on them and because of them. No longer are we worried about what a smartphone can really do; the basics are there and incremental changes don't mean as much anymore. But the potential of the devices through software and technologies like GPS means they still have much more to offer us because of where they are in terms of development.

We are living through times where people who have never been in a classroom can have more work experience than the average college graduate. Tapping into new skillsets and smart people will require new techniques, open minds and new procedures. This is, of course, if they want to work with you. Fresh talent, political uprisings, open minds, new systems and old systems will all play a part in the rich tapestry of disruption. There are more disruptive technologies on the way than we have discussed here, and I for one look forward to seeing what world we live in when the following really begin to shape aspects of our lives:

- *Next-generation batteries.* Dyson, Musk and others are bullish on renewable energy and now that supply is meeting demand thanks to lower costs and energy storage improvements, new 'mini grids' are possible which will disrupt power companies and possibly whole countries. Beyond powering homes and offices, supercapacitors charge faster and are more flexible than other existing options – perfect for handheld and small devices.

- *Nanosensors.* There are billions of connected devices but where this area of technology gets interesting is with healthcare and buildings. Having sensors inside the body, or houses that can alert and fix issues, will disrupt multiple markets and companies to innovate and provide new services and solutions.

- *Autonomous vehicles.* The dream of self-driving vehicles for many meets with visions from Hollywood but others are thinking about the future of the office, the systems these vehicles could create and what this means for roads and other service industries. Uber, Tesla, Google and most car manufacturers are heavily invested in bringing this vision to fruition sooner rather than later to stave off disruption and corner markets.

- *Quantum computing.* Quantum computers have the potential to perform certain calculations significantly faster than any silicon-based computer. Recent improvements to the lifetime of superconducting quantum circuits mean that harnessing the power of atoms and molecules to perform memory and processing tasks is closer to reality than ever before. Microsoft and Google are already racing to make these computers to gain a competitive edge and offer new services that they can monetize.

Most if not all of these technologies have two big things in common: one, they are building on existing technologies to create new systems of value, and two, they deal directly with value to the individual – an important distinction with some of the technologies covered in this book. Some of these are brand-new technologies and the full effects of them have yet to be imagined, let alone seen. The connection to value to the individual will remain key for successful disruptive technologies. The technologies that will most disrupt the future will always have a significant benefit to human life and wellbeing; great examples are robotic exoskeletons and drone technology. Both of these examples show the pros and cons side of technologies well. On one hand you have a suit that will help people lift heavy items and have a better standard of living, but you also have a suit that can throw heavy objects at others. Similarly, drones offer a huge benefit to humanity, from aerial surveying to delivery. Equally, drones have been used to target individuals for surveillance. However, though technology isn't inherently evil, the potential for misuse is always there, hence the B in TBD must always be considered carefully.

The future of TBD

TBD is not a static procedure and I have been thinking hard recently about steps to alter it to include something I am passionate about – design. Design is something I have always been interested in; when form and function truly work together, magic happens.

Design is a subject as big as it is subjective but recently I have seen more and more written about it as design thinking takes off and more tactile devices flood markets. Whether it is data visualization, user experience, user interaction, usability, interaction design, visual design, systems design or the myriad other sub-genres that are springing up, design is increasingly important.

Consumers demand that things work first time, and are easy to use and intuitive. Companies have never had it harder (or easier) to please the final user but still we seem bad at doing just that.

Previously, business has been defined by war metaphors and terminology but those days are numbered because of the new wave of collaboration, learning and value-based outcomes we are beginning to see in businesses like Unilever, Google and the start-up culture in general.

As we head towards 8 billion mobile users (currently there are around 7.1 billion) mixed with Google, Amazon, Meta and Apple (GAMA for short;

you'll start to hear this more and more frequently now) – tech giants who make their own hardware – unprecedented opportunities are arising for brands to essentially build on the shoulders of giants (a phrase made popular by Isaac Newton). New hardware opportunities, new networks to exploit and utilize, new markets to uncover and new content to create – there are more ways to compete than ever before. Now factor in the recent developments of machine learning (AI) and we get down to single-digit error rates for things like speech recognition and visual recognition. An incredibly interesting future is emerging that mixes computers, humans and super-fast calculations – or technology, behaviour and data. The only variable is you. What impact will you have? What will you add to the future? How bold will you be, or allow others to be? How frictionless will you make your business in one, two, five years?

Never before have businesses had a bigger opportunity to change and create amazing products and services that not only make people's lives better but indeed change the very world we live in – or perhaps even lead to a new one. The pandemic, while destructive and harrowing, has handed multiple opportunites to most, if not all, businesses who can identify and seize them.

The world we live in is changing, physically, spiritually and technologically. Businesses are about to experience the need to respond to scenarios and climates that have never existed before. Agencies will be making things they've never made before and the coming decade will require agencies and businesses to start to think about things like virtual reality rights, identity brokers, smart contact developers, waste data managers and mixed reality architects. Increased collaboration will create friction as much as it will create opportunities and push boundaries. Existing and new networks around things like wearables will spring up and be gone overnight but each will offer an opportunity to people who are prepared, agile and flexible enough. Experts will come and go but ultimately those who are adaptive and think critically will see the most success because, honestly, a lot of the future will not have existed before in any shape or form. You can fear it or you can prepare for it.

Short-termism continues to be something I hear being discussed with increasing frequency at industry conferences but worryingly little in the business and agency boardrooms I frequent. In every region, there is uncertainty that seems to be growing despite the availability of more data and smarter tools. I urge you to use this book as the start of your permission to stop thinking in the short term. Force yourself to think ahead and make plans – plans can always change but if your north star is low, while easier to reach, it is unlikely to satisfy or create value for anyone. The best goals force you to reach further – be bold. With people at all levels of an organization staying in positions more briefly than ever, along with a generally challenging landscape, going for the easy win or the short-term goal is desirable but this is not where success or greatness lies. Both these things will always be given to those who push harder and higher than the rest. There is no participation award when it comes to greatness and disruption. The time to take risks is never next year. Plan and ruthlessly execute – make a conscious decision to think long-term using TBD and TBD+.

Disruptive Technologies started out with a look backwards to see where technologies came from and why they were created. As the book progressed we saw how, more and more, factors beyond simple technological advances – like behavioural, economic and political disruption – become factors in future technological advancements; disruption seems to be coming from all sides. The next period of technological advancement is critical for the human race because of the systems provided by the last period of technology, like social networks, live broadcast, sensor networks and GPS to name a few. As GAMA – and other disruptive powerhouses – continue to morph from their original businesses, new, exciting and scary propositions emerge. An open but critical mindset is required in order to see the possibilities and understand the consequences of the choices being made by these companies, and so thrive on the disruption they will cause. Whatever continent you reside on, it is clear we are all living through incredibly uncertain times thanks in large part to a changing class system, changing political landscape and economic upheaval. There has never been a more important time to have a clear perspective and strategy for handling disruptive technologies.

Thank you for reading – you'll never have a better time to make changes than right now – you've already started to push your company to new heights. Don't stop now, keep going.

REFERENCES

Adams, D (2002) *The Salmon of Doubt*, Heinemann

Beckhard, R (1975) Strategies for large system change, *Sloan Management Review*, **16** (2)

Bodell, L (2012) *Kill the Company: End the status quo, start an innovation revolution*, Bibliomotion

Bower, J (2002) Disruptive change, *Harvard Business Review*, 80 (5), pp 95–101

Bower, J and Christensen, C (1995) Disruptive technologies: catching the wave, *Harvard Business Review*, 73 (1), pp 43–53

Burgess, C (2016) Interview with author, 12 July

Cady, S H, Jacobs, J, Koller, R and Spalding, J (2014) The change formula: myth, legend, or lore, *OD Practitioner*, **46** (3)

CEB (2016) 'Risk management' is often synonymous with 'risk prevention' but it shouldn't be, Cebglobal, www.cebglobal.com/risk-audit/risk-management/how-to-live-with-risks.html (archived at https://perma.cc/W7VL-TEE9)

Clegg, N (2022) Making the metaverse: What it is, how it will be built, and why it matters, https://nickclegg.medium.com/making-the-metaverse-what-it-is-how-it-will-be-built-and-why-it-matters-3710f7570b04 (archived at https://perma.cc/MBD4-NCDE)

Dearborn, J (2015) *Data Driven*, John Wiley & Sons

Diamandis, P and Kotler, S (2015) *Bold: How to go big, create wealth, and impact the world*, Simon & Schuster International

Diebold, F (2012) A personal perspective on the origins and development of 'Big Data': the phenomenon, the term, and the discipline, University of Pennsylvania, www.ssc.upenn.edu/~fdiebold/papers/paper112/Diebold_Big_Data.pdf (archived at https://perma.cc/D33G-UERR)

Diehl, S (2022) Web3 is Bullshit, Stephendiehl.com, www.stephendiehl.com/blog/web3-bullshit.html (archived at https://perma.cc/MTT6-9G8S)

Dorsey, J [jack] (2021) [You don't own "web3." The VCs and their LPs do. It will never escape their incentives. It's ultimately a centralized entity with a different label. Know what you're getting into...] 21 December, https://twitter.com/jack/status/1473139010197508098?s=20&t=kN9hRJHFrXeOXRD5JbEkRw (archived at https://perma.cc/Y8XA-5ZUR)

Everything is a Remix (2015) http://everythingisaremix.info/watch-the-series/ (archived at https://perma.cc/7GXK-GS8C)

Finney, H (2006) Quiz: fox or hedgehog? Overcoming Bias, www.overcomingbias.com/2006/11/quiz_fox_or_hed.html (archived at https://perma.cc/9RRM-55JG)

Franke, N, Poetz, M and Schreier, M (2011) The value of crowdsourcing: can users really compete with professionals in generating new product ideas? *Journal of Product Innovation Management*, **29** (2), pp 245–56

Gardner, J (1990) Personal renewal, PBS, www.pbs.org/johngardner/sections/writings_speech_1.html (archived at https://perma.cc/ARQ9-MLA5)

Goodwin, T (2015) The battle is for the customer interface, TechCrunch, http://techcrunch.com/2015/03/03/in-the-age-of-disintermediation-the-battle-is-all-for-the-customer-interface/#.f2ueb6:0sCd (archived at https://perma.cc/NWD4-M5HH)

Goyder, C (2014) The surprising secret to speaking with confidence (TED talk), YouTube, https://youtu.be/a2MR5XbJtXU (archived at https://perma.cc/T72D-77T5) [accessed 20 July 2022]

Hackman, R (1973) *Group Influences on Individuals in Organizations*, Yale University Press

India Daily Star (2012) Apply nanotech to up industrial, agri output, http://archive.thedailystar.net/newDesign/news-details.php?nid=230436 (archived at https://perma.cc/9NNP-DC5Y)

Intel (2015) 50 years of Moore's Law, www.intel.com/content/www/us/en/silicon-innovations/moores-law-technology.html (archived at https://perma.cc/K3RD-MLBX)

Ioannou, L (2014) A decade to mass extinction event in S&P 500, CNBC, www.cnbc.com/2014/06/04/15-years-to-extinction-sp-500-companies.html (archived at https://perma.cc/8ZGJ-KR64)

Jones, B (2005) Age and great invention, National Bureau of Economic Research, www.nber.org/papers/w11359 (archived at https://perma.cc/AK7X-335E)

Keldsen, D (2016) Interview with author, 19 April

Kelley, T (2008) *The Ten Faces of Innovation: Strategies for heightening creativity*, Profile Books

Kim, H, Park, J, Cha, M and Jeong, J (2015) The effect of bad news and CEO apology of corporate on user responses in social media, *PLOS ONE*, **10** (5)

Lux Research (2014) Nanotechnology update: corporations up their spending as revenues for nano-enabled products increase, portal.luxresearchinc.com/research/report_excerpt/16215 (archived at https://perma.cc/T2NV-MNQ6)

Maeda, J (2006) *The Laws of Simplicity: Design technology, business, life*, The MIT Press

Maslow, A (1943) A theory of human motivation, *Psychological Review*, 50, pp 370–96

Norton, S (2015) Internet of Things market to reach $1.7 trillion by 2020, *The Wall Street Journal*, http://blogs.wsj.com/cio/2015/06/02/internet-of-things-market-to-reach-1-7-trillion-by-2020-idc/ (archived at https://perma.cc/MM8K-VRSL)

Olson, D (2022) Line Goes Up – The Problem With NFTs, www.youtube.com/watch?v=YQ_xWvX1n9g&ab_channel=FoldingIdeas (archived at https://perma.cc/9XLB-GSJV)

Pontin, J (2015) Interview with author, 8 November

Poswolsky, A (2016) Interview with author, 12 July

Rogers, E (2003) *Diffusion of Innovations*, 5th edn, Free Press

Rotolo, D, Hicks, D and Martin, B (2014) What is an emerging technology? *SSRN Electronic Journal*, https://papers.ssrn.com/sol3/papers.cfm?abstract_id=2564094 (archived at https://perma.cc/LL8H-MV73)

Russon, M-A (2016) Interview with author, 13 July

Sanburn, J (2015) How every generation of the last century got its nickname, *Time*, http://time.com/4131982/generations-names-millennials-founders/ (archived at https://perma.cc/77Q5-UGPW)

Shedden, D (2014) Today in media history: Mr Dooley: 'The job of the newspaper is to comfort the afflicted and afflict the comfortable', Poynter, www.poynter.org/2014/today-in-media-history-mr-dooley-the-job-of-the-newspaper-is-to-comfort-the-afflicted-and-afflict-the-comfortable/273081/ (archived at https://perma.cc/W5AD-V49M)

Sinek, S (2011) *Start with Why: How great leaders inspire everyone to take action*, Portfolio

Sinek, S (2017) Interview with author, 4 January

Smith, R (2009) Nanoparticles used in paint could kill, research suggests, *Daily Telegraph*, www.telegraph.co.uk/news/health/news/6016639/Nanoparticles-used-in-paint-could-kill-research-suggests.html (archived at https://perma.cc/H2Q4-DSKU)

Stephenson, N (1982) *Snow Crash*, Bantam Books

TBD (n.d.) www.tbd.website/abo (archived at https://perma.cc/J4CQ-LETC)

Tetlock, P E (2015) *Expert Political Judgment*, Princeton University Press

Tetlock, P E and Gardner, D (2016) *Superforecasting: The art and science of prediction,* Random House

Tett, G (2015) *The Silo Effect: The peril of expertise and the promise of breaking down barriers*, Simon & Schuster

Trzesniewsk, K and Donnellan, M (2010) Rethinking 'Generation Me': a study of cohort effects from 1976–2006, *Perspective on Psychological Science*, 5 (1), pp 58–75

US Chamber of Commerce Foundation (2012) The Millennial Generation Research Review, www.uschamberfoundation.org/reports/millennial-generation-research-review (archived at https://perma.cc/UM2F-B2G8)

US Department of Defense (2002) DoD new briefing: Secretary Rumsfeld and Gen Myers

Vanderkam, L (2011) *168 Hours: You have more time than you think*, Portfolio

Vanderkam, L (2015) How to do your own time makeover, http://lauravanderkam.com/wp-content/uploads/2013/05/How-To-Do-Your-Own-Time-Makeover.pdf (archived at https://perma.cc/394G-W54F)

Wohlers (2015) Wohlers Report 2014, www.wohlersassociates.com/2015report.htm (archived at https://perma.cc/C9B7-ZZ3B)

FURTHER READING

Here is a collection of books I have read and return to for inspiration. I recommend you read them.

Chomsky, N, Barsamian, D and Naiman, A (2012) *How the World Works*, Hamish Hamilton

Cialdini, R B (2007) *Influence: The psychology of persuasion*, HarperCollins

Diamandis, P H and Kotler, S (2015) *Bold: How to go big, create wealth and impact the world*, Simon & Schuster

Dobbs, R, Manyika, J and Woetzel, J (2015) *No Ordinary Disruption: The four global forces breaking all the trends*, PublicAffairs

Duncan, K (2013) *The Diagrams Book: 50 ways to solve any problem visually*, LID Publishing

Duncan, K (2014) *The Ideas Book: 50 ways to generate ideas more effectively*, LID Publishing

Eggers, D (2014) *The Circle: A novel*, Hamish Hamilton

Gneezy, U, List, J A and Levitt, S D (2013) *The Why Axis: Hidden motives and the undiscovered economics of everyday life*, PublicAffairs

Godin, S (2012) *The Icarus Deception: How high will you fly?* Portfolio Penguin

Goyder, C (2015) *Gravitas: Communicate with confidence, influence and authority*, Random House

Heath, C and Heath, D (2011) *Switch: How to change things when change is hard*, Random House Business Books

Keldsen, D (2014) *The Gen Z Effect*, Bibliomotion

Kelley, T and Littman, J (2008) *The Ten Faces of Innovation: Strategies for heightening creativity*, Profile Business

Klein, G (2014) *Seeing What Others Don't: The remarkable ways we gain insights*, Nicholas Brealey Publishing

Krogerus, M and Tschäppeler, R (2010) *The Decision Book: Fifty models for strategic thinking*, Profile Books

Leaf, R (2012) *The Art of Perception: Memoirs of a life in PR*, Atlantic Books

Levitt, S D and Dubner, S J (2009) *Superfreakonomics: Global cooling, patriotic prostitutes and why suicide bombers should buy life insurance*, Penguin Books

Maeda, J (2006) *The Laws of Simplicity: Design, technology, business, life*, The MIT Press

Mason, H, Mattin, D and Luthy, M (2015) *Trend-driven Innovation: Beat accelerating customer expectations*, John Wiley & Sons

Nabben, J (2013) *Influence: What it really means and how to make it work for you*, Pearson Education

Nisbett, R 2015) *Mindware: Tools for smart thinking*, Allen Lane

Orwell, G (2013; 1949) *Nineteen Eighty-Four*, Penguin Classics

Outram, C (2015) *Digital Stractics: Where strategy and tactics meet and bin the strategic plan?* Palgrave Macmillan

Pink, D H (2014) *To Sell is Human: The surprising truth about persuading, convincing, and influencing others*, Canongate Books

Poswolsky, A (2016) *The Quarter-Life Breakthrough: Invent your own path, find meaningful work, and build a life that matters*, TarcherPerigee/Penguin Random House

Quartz, S and Asp, A (2015) *Cool: How the brain's hidden quest for cool drives our economy and shapes our world*, Farrar, Straus and Giroux

Rose, J (2016) *Flip the Switch: Achieve extraordinary things with simple changes to how you think*, Capstone Publishing

Saatchi, M (2013) *Brutal Simplicity of Thought: How it changed the world*, Ebury Press

Tetlock, P and Gardner, D (2015) *Superforecasting: The art and science of prediction*, Random House Books

Vlaskovits, P, Koffler, J and Patel, N (2016) *Hustle: The power to charge your life with money, meaning and momentum*, Vermilion

INDEX

CPSIA information can be obtained
at www.ICGtesting.com
Printed in the USA
BVHW011809060223
657988BV00011B/133